INTERNATIONAL SERIES OF MONOGRAPHS ON

ELECTRONICS AND INSTRUMENTATION

GENERAL EDITORS: D. W. FRY and W. HIGINBOTHAM

VOLUME 19

OTHER TITLES PUBLISHED IN THE SERIES

(FORMERLY PERGAMON SCIENCE SERIES ELECTRONICS AND WAVES)

Vol. 1 *Signal Noise and Resolution in Nuclear Counter Amplifiers*
by A. B. GILLESPIE

Vol. 2 *Scintillation Counters*
by J. B. BIRKS

Vol. 3 *Probability and Information Theory with Application to Radar*
by P. M. WOODWARD

Vol. 4 *Physics and Applications of Secondary Electron Emission*
by H. BRUINING

Vol. 5 *Millimicrosecond Pulse Techniques (2nd edition)*
by I. A. D. LEWIS and F. H. WELLS

Vol. 6 *Introduction to Electronic Analogue Computers*
by C. A. A. WASS

Vol. 7 *Scattering and Diffraction of Radio Waves*
by J. R. MENTZER

Vol. 8 *Space-charge Waves and Slow Electromagnetic Waves*
by A. H. W. BECK

Vol. 9 *Statistical Theory of Signal Detection*
by CARL W. HELSTROM

Vol. 10 *Laplace Transforms for Electronic Engineers*
by J. G. HOLBROOK

Vol. 11 *Frequency Modulation Theory—Application to Microwave Beams*
by J. FAGOT and PH. MAGNE

Vol. 12 *Theory of Microwave Valves*
by S. D. GVOZDOVER

Vol. 13 *Electronic Computers*
by A. I. KITOV and N. A. KRINITSKII

Vol. 14 *Topics in Engineering Logic*
by M. NADLER

Vol. 15 *Environmental Testing Techniques*
by G. W. A. DUMMER and N. B. GRIFFIN

Vol. 16 *Fundamentals of Microwave Electronics*
by V. N. SHEVCHIK

Vol. 17 *Static Electromagnetic Frequency Changers*
by L. L. ROZHANSKII

Vol. 18 *Problems in the Design and Development of 750 MW Turbogenerators*
by V. P. ANEMPODISTOV, E. G. KASHARSKII and I. D. URUSOV

CONTROLLED-DELAY DEVICES

CONTROLLED - DELAY DEVICES

by

S. A. DOGANOVSKII and V. A. IVANOV

Translated from the Russian by

O. M. BLUNN

Translation edited by

D. K. GHOSH

A Pergamon Press Book

THE MACMILLAN COMPANY

NEW YORK

1963

THE MACMILLAN COMPANY
60 Fifth Avenue
New York 11, N.Y.

This book is distributed by
THE MACMILLAN COMPANY
pursuant to a special arrangement with
PERGAMON PRESS LIMITED
Oxford, England

This translation has been made from a book by S. A. Doganovskii and V. A. Ivanov entitled *Bloki Reguliruyemogo Zapazdyvaniya* published by Gosenergoizdat, Moscow, 1960.

Library of Congress Card Number 63–10118

Printed in Great Britain by
Eynsham Technical Press (Oxford) Ltd.

ANNOTATION

This book describes devices which can produce a pure time lag in control and computer engineering.

Their construction and circuits are considered and practical recommendations are made concerning their application.

The book is written not only for professional control engineers, but for anyone concerned with the development or utilisation of controlled-delay devices in industry.

CONTENTS

PREFACE

Devices which produce a pure time delay are extensively used in solving many of the problems associated with the automation of production and the design of control systems. They are also used in the experimental investigation of control systems by analogue techniques and as individual elements in systems of automatic control and regulation.

Controlled-delay devices are delay devices in which the time lag can be regulated by external control signals. Though these are most widely used, they are not as yet produced on an industrial basis. The various organizations which use them have had to develop and design them for themselves. What little published literature there is on this topic, as a few articles in journals, contains little information about the circuits of controlled-delay devices or the most suitable device for a particular application.

This book aims to fill this gap. It can also be used as a manual on controlled-delay devices.

The Authors

CHAPTER I

1.1 Applications and Basic Principles of Controlled-Delay Devices

In many technological processes the raw material can be regarded as moving linearly in space between different stages of production. Such is the case for rolling mills, calendering lines and paper-making machines, where the material moves along a single line through various stages of production. In chemical engineering, liquids are transported in pipes from one unit to another and so on.

To devise an automatic control system for such processes, or to simulate them mathematically, it is very important to obtain timely information about those changes in the properties of the moving material which can only be measured, for example, at the beginning of the process. For a continuous rolling mill, it is necessary to know the thickness of the material as it enters the last stands of the mill, even though there is only one device for measuring the thickness of the billet at the input to the mill (or after the first stand).

A similar problem arises on calender lines and in paper-making machinery. In chemical production, it is necessary to determine the instant at which a substance with known properties reaches a unit where it is mixed with some reagent and so on.

The production processes in which such delays occur cannot be provided with a modern efficient system of automatic control or studied by present-day simulation techniques, unless special devices are available for storing essential information about the properties of the moving material for the required length of time.

Controlled-delay devices serve this purpose. Their output signal reproduces an input signal after a specified period of time which has been automatically matched with the characteristics of the production process (e.g. a variation of strip speed on calender lines or paper-making machinery).

The main fields of application of controlled-delay devices are:

1. in simulation of automatic control systems with time lag,
2. in optimalizing control systems (as elements of compensating devices),
3. in automatic inspection systems where instantaneous data-storage is required,
4. in analysis and synthesis of automatic control systems.

A pure delay τ is above all required in the simulation of industrial control systems.

Industrial plants which are subject to several controls are described by high order equations which can be replaced by first and second order equations with a pure delay. For some industrial plants the time lag is due to the finite rate of transfer of the material. Consequently, in simulating such plants, it is necessary to have a device which produces a pure time lag τ.

The time lag for the various control systems and processes can range from a fraction of a second to tens of minutes. The corresponding frequency band ranges from zero to tens of cycles per second.

A delay device must produce the time lag within the prescribed time limits and satisfy requirements as regards the transmission of the required frequency band width.

The relationship between the input signal $U_{in}(t)$ of the delay device and its output signal $U_{out}(t)$ in the ideal case can be represented as

$$U_{out}(t) = U_{in}(t - \tau),$$

where τ is the delay time. The output quantity $U_{out}(t)$ must therefore equal the input $U_{in}(t)$, but after a delay time τ.

To find the transfer function of a delay device, it is convenient to re-write the above equation in operator form:

$$U_{out}(p) = e^{-p\tau} U_{in}(p).$$

The transfer function, which is the ratio of the transforms of the output and input quantities, is then

$$W'(p) = e^{-p\tau}.$$

An ideal delay device would have an amplitude-frequency characteristic $|W'(j\omega)| = 1$, and a phase-frequency characteristic $\varphi' = \omega\tau$. But in real delay devices these two characteristics cannot be reproduced exactly at the same time.

The systems so far used, only permit exact reproduction of the amplitude-frequency characteristic and approximate reproduction of the phase characteristic, or vice versa, exact reproduction of the phase characteristic and approximate reproduction of the amplitude-frequency characteristic.

The simplest delay devices are electrical circuits which are composed of series-connected passive and active four-pole networks. Such systems only have a limited range of fixed delay times and they are therefore little suited for use in controlled-delay devices.

Systems which in principle reproduce the phase-frequency characteristic exactly, and the amplitude-frequency characteristic approximately, require delay devices using storage elements. In these devices, the instantaneous values of the input are continuously stored and then reproduced or sampled after the required time lag τ.

Tape recorders and capacitors are the most widely used storage units. Here the time lag can be varied by changing the speed of the tape or that of the brush which samples the voltage from the capacitors. This speed and, therefore, the delay can be set by a servo system with external control signals.

CHAPTER II

MAGNETIC TAPE CONTROLLED-DELAY DEVICES

2.1 Design Method and the Principal Units

Tape recorders are widely used in the construction of controlled-delay devices. This system allows wide variation of the time delay and permits transmission of a sufficiently wide frequency band of signals. For the transmission of signals having a constant component and ultra-low frequencies, especially in process control systems, it is first necessary to modulate a certain carrier frequency by these signals. It is then possible to produce a pure delay for processes with frequencies from zero to 100-200 c/s.

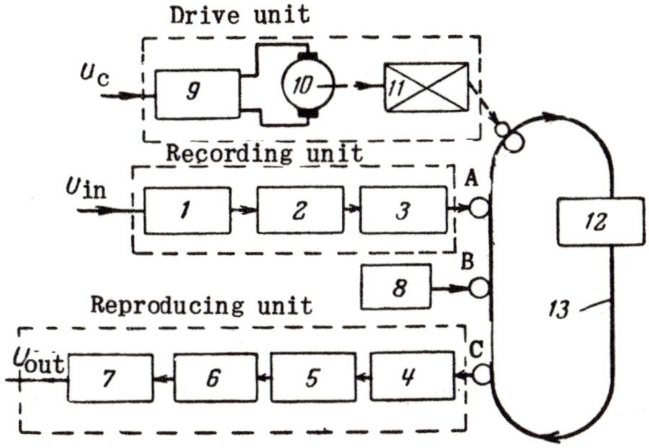

Fig. 1. Block diagram of a controlled-delay device using magnetic recording:

1 - input unit; *2* - frequency modulator; *3* - power amplifier; *4* - amplifier; *5* - limiter; *6* - frequency demodulator; *7* - output unit; *8* - supply source; *9* - speed control system of motor; *10* - motor; *11* - reduction gears; *12* - continuous reel; *13* - magnetic tape; *A* - recording head; *B* - erase head; *C* - playback head.

Devices using magnetic recording are illustrated in Fig.1. The input signal U_{in} is fed to the recording head via an input device, modulator and power amplifier and then recorded on the tape. After a specified period of time, which depends on the length of tape between the recording and playback heads and the speed of the tape, the recorded signal induces an e.m.f. in the playback head which is then amplified. After passing through the demodulator the output is made equal to the input voltage, but delayed in time by τ. The erasing head prepares the tape for the next cycle. To use tapes of up to tens and even hundreds of metres in length between the heads for large time lags, use may be made of a "continuous reel" which maintains the motion of the tape constant. This special attachment will be considered again later.

The circuits of the individual elements of the delay device (the modulator, demodulator etc.), depend on the type of modulation used for recording input voltages of ultra-low frequency. In principle, amplitude-, frequency- and pulse-modulation can be used.

In amplitude modulation, the amplitude envelope of the carrier frequency oscillations varies according to the law governing the variation of the input signal. Since the input signal is sign-variable, it is necessary to have an initial displacement of the carrier oscillation. In some cases, the change of sign is taken into account by a change in phase or a change in carrier frequency. The carrier frequency ω_0 is selected within the most favourable limits for magnetic recording (of the order of 500-5000 c/s). The width of the frequency spectrum of the signal which is to be recorded, is twice the frequency of the input signal Ω and is confined to the range $(\omega_0-\Omega)$— $(\omega_0+\Omega)$. In recording signals at ultra-low frequencies ($\Omega_{max} \leqslant 100$ c/s, the band width is comparatively narrow. This is the advantage of amplitude modulation. Systems which effect amplitude, modulation and demodulation are also very simple. This is yet another advantage of amplitude modulation. But in employing amplitude modulation, the accuracy of the device is greatly affected in operation by the internal noise of the recording apparatus, due mainly to the non-uniformity of the tape over its length because of the different size of the individual crystals of the magnetic tape and their arrangement.

The geometric dimensions of magnetic coatings cannot be maintained with absolute accuracy. As a result, with non-uniform thickness and width of the coating, a similar recording head field will produce different residual magnetization of the tape.

When using standard annular heads for recording and playback, the gap between the tape and core of the head is continually changing owing to the irregular surface of the tape and the vibration of the tape when in motion.

In practice, the amount of noise is such that a recording signal of constant amplitude will vary after reproduction by ± 1 to 1.5 dB. The error in the reproduction of amplitude-modulated signals is therefore 12 to 18 per cent when using standard apparatus. This error can be reduced, to 5 to 7 per cent by adding special attachments to keep the gap between the tape and heads constant. This high level of noise and the non-linearity of the recording characteristics can considerably reduce the dynamic range of a device to 25-30 dB. The best results from the point of view of the dynamic range are obtained by using amplitude modulation with suppression of the carrier. But the error in reproduction due to non-uniformity in the sensitivity of the tape still remains. These disadvantages prevent the use of amplitude modulation in controlled-delay devices.

With frequency modulation, the oscillation amplitude remains constant, but the frequency of the oscillations is varied in accordance with the amplitude of the modulating signal. The percentage modulation factor ν is in this case given by the ratio $\nu = \frac{\Delta\omega}{\omega_0}$, where $\Delta\omega$ is the frequency deviation and ω_0 the carrier frequency.

In reproducing frequency-modulated oscillations, the signal from the playback head is amplified and then fed to a limiter where variations in amplitude of the modulated oscillations due to tape non-uniformity and internal noise are eliminated. This amplitude limiting makes it possible to reproduce ultra-low-frequency signals with a very high degree of accuracy. The main source of error in frequency modulation is the fluctuation in the tape speed due to inaccuracies in the production of the driving unit of the tape-tensioning device and variation in the speed of the drive motor. This fluctuation produces additional variations in the frequency of the recorded signal which results in error during reproduction. The value of error depends on the modulation factor ν and the variations in speed. For 50 per cent modulation ($\nu = 0.5$) and 0.2 per cent speed variation, which is easily attainable in simple systems, the maximum fluctuation-induced error in the performance of a device using frequency modulation is only 0.6 per cent.

The carrier frequency needs to be selected, so that the spectrum of the frequency modulated oscillations is in the frequency band width (200-5000 c/s) which has the minimum attenuation in transmission at relatively low tape speeds.

With frequency modulation the non-linearities of magnetic recording do not affect the accuracy with which a low frequency signal is transmitted and it is therefore possible to dispense with ultrasonic bias in recording and erase the recording by direct current. This considerably simplifies the recording and erasing units. The error in reproduction is due to variation of the tape speed and, consequently, the fluctuations of frequency can be compensated by special techniques to improve the performance of the whole delay device.

With pulse modulation, it is advisable to use pulse-width modulation or pulse-code modulation in recording units for ultra-low frequency signals. When using these types of modulation, the same as in frequency modulation, the accuracy of the device is unaffected in operation by spurious amplitude modulation, internal noise, or non-linearity in the characteristics of the tape. The use of pulse-width and pulse-code modulation in some cases gives greater accuracy in the performance of the recording unit than frequency modulation. But the recording and playback units are rather more complicated.

2.2 Recording Unit

The recording unit of a delay device contains an input unit *1*, a frequency modulator *2* and a power amplifier *3* (Fig. 1). The purpose of the input unit *1* is to match the parameters of the signal source with those of the following circuits in the recording unit. After passing through the input unit, the signal is fed to a frequency modulator which produces a voltage at a frequency which is proportional to the amplitude of the input signal.

The main types of circuit for a frequency modulator are:

1. a positive bias multivibrator;
2. a transitron oscillator;
3. a saw-tooth oscillator with a capacitor discharging through a valve.

In the positive bias multivibrator, the leakage resistances of the grids are connected to a positive voltage source. The frequency at the output of such a multivibrator is a linear function of the positive voltage applied to the grids. That is why it can be used as a frequency modulator. The degree of linearity of the characteristic which expresses the relationship between the frequency of the multivibrator and the valve grid voltage is sufficiently high over a wide band of frequencies that a large percentage modulation factor (± 50 per cent) can be obtained. The advantage of such modulators is that they are simple to make and service. The drawback is the large input voltage which is required (± 70 to 100 V).

The fundamental circuit of a typical recording unit in which a positive bias multivibrator is used as the modulator is shown in Fig. 2, *a*. The initial positive voltage on the grids of the valve V_2 of the multivibrator is provided by an amplifier (valve V_1) with a comparatively low but stable gain owing to considerable negative feedback. The value of the gain depends solely on the ratio of the resistances R_1/R_3.

The frequency of the multivibrator depends on the quantities C_2, C_3, R_6 and R_7 and the ratio of the voltages U_m/E_a (where U_m is the multivibrator voltage and E_a the anode supply voltage). On variation of the input signal, the voltage U_m varies and so does the frequency of the voltage at the output of the multivibrator, which, via an amplifier (the valve V_3), enters the recording head and is recorded on the tape.

For the parameter values indicated in Fig.2, the characteristic $f_{out} = \varphi(U_{in})$ has a frequency deviation $\Delta f = \pm$ $\pm$ 600 c/s for $f_o = 1200$ c/s and $U_{mo} = 150$ V. In this case, the voltage U_m varies between 40 and 260 V.

To ensure that the voltage in the recording head is rectangular in shape, a transformer Tr_1 is set up at the output of the power amplifier. Its primary winding is the load of a cathode follower. The recording head is connected to its secondary winding.

The transitron oscillator system allows the stated modulation factor to be obtained at a comparatively low input voltage (± 10 V), but its characteristic is less linear with the result that the accuracy of the device in operation is reduced. The fundamental circuit of a typical recording unit using this system is shown in Fig. 2, *b*. The input signal is fed through a cathode follower V_1 to the grid of the valve of the V_2 from the anode of which a saw-tooth voltage is taken. The frequency of this voltage varies as a function of the grid bias of the valve V_2, which is determined by the input voltage.

For the indicated parameter values, the characteristic $f_{out} = \varphi(U_{in})$ has a frequency deviation $\Delta f = \pm 1800$ c/s for $f_o = 3000$ c/s. In this case, the variation of the input voltage is ± 10 V.

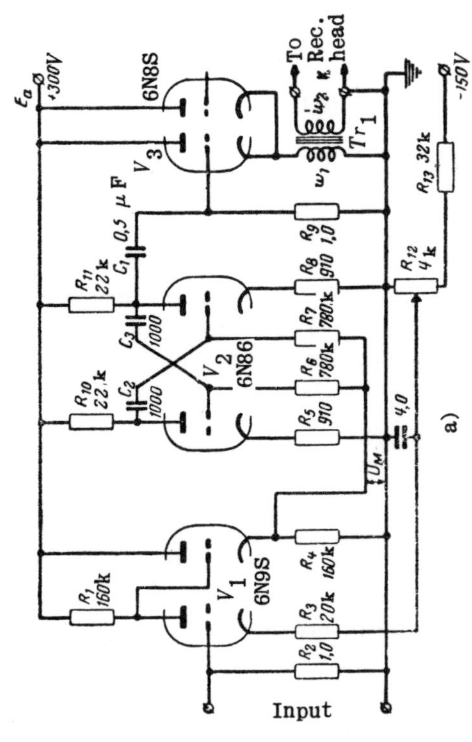

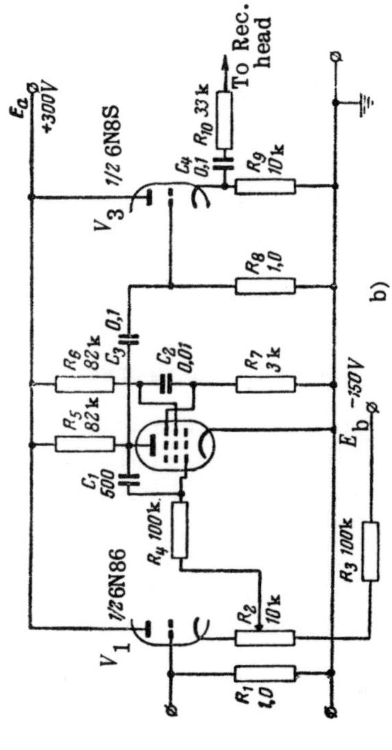

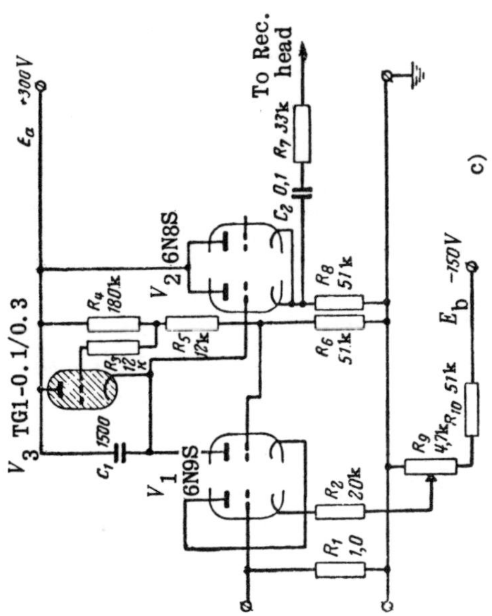

Fig. 2. Circuit diagrams of recording units:

a - with a multivibrator; *b* - with a transitron oscillator; *c* - with a saw-tooth oscillator.

Figure 2, *c* shows the circuit of a typical recording unit in which a saw-tooth oscillator is used as the modulator. This oscillator is analogous to the circuit used for producing the saw-tooth voltage of cathode-ray oscillographs, except that a double triode is used instead of a pentode as the charging valve to increase the slope of the modulation characteristic.

This system permits a smaller modulation factor (± 40 per cent), than that with a multivibrator, but little change is required in the input voltage ± 4 V.

The frequency of the oscillator depends on the charging current of the capacitor C_1 through the valve V_1 and the value of the capacitance C_1, and is obviously proportional to the input signal provided that there is a linear relationship between the input signal and the valve current. Since the modulator has a high input impedance and requires no bias voltage, the input unit *1* is not required.

The choice of recording unit for each particular case depends on the requirements which the delay device must meet as regards linearity, operating accuracy, frequency deviation, the shape of the carrier frequency voltage and so on.

2.3 Playback Unit

When the tape with the signal recorded on it moves past the playback head, an e.m.f. is induced in this head and the magnitude of this e.m.f. depends on the speed of the tape and the frequency of the recorded signal. Since this induced e.m.f. is very small, it is usually necessary for the signal to be amplified before feeding it to the frequency demodulator. The frequency demodulator may or may not incorporate an additional time delay device.

In the system with incorporated time delay, the signal from the reproducing head is fed after amplification to a limiter and after limiting it, is differentiated. The pulses thus formed serve to trigger a multivibrator with one stable state. This multivibrator produces a train of rectangular pulses of strictly constant magnitude and width. The mark-space-ratio (the interval between pulses) is determined by the frequency of the recorded signal or, in the last analysis, by the amplitude of the input signal. When this "transformed" signal has been transmitted through a conventional detector (rectifier) and low pass filter, a voltage is obtained which is identical to the input signal voltage in magnitude. The output voltage of such a system can be comparatively large (several tens of volts) with good linearity. That is the main advantage of the system. but the system is complicated and there is an increased error due to the two temporal transformations.

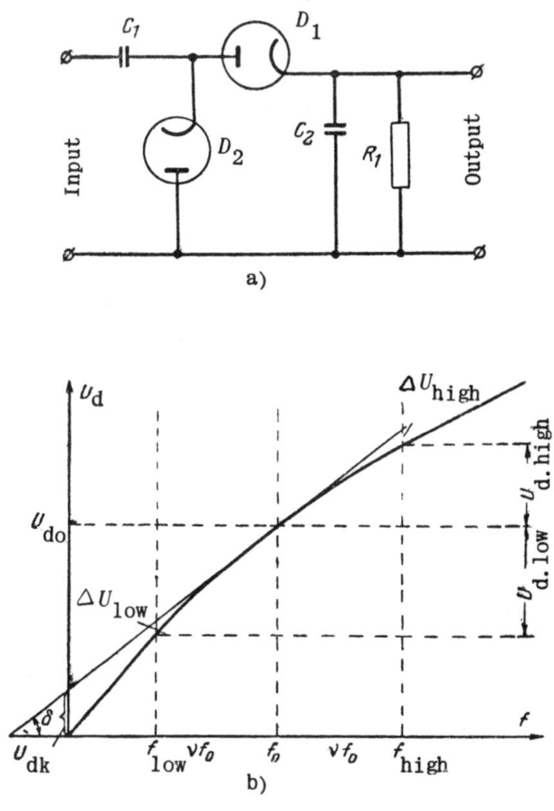

Fig. 3. Diode detector:

a - circuit diagram; *b* - the output voltage as a function of the frequency at the input.

In the simple frequency demodulator system without two time conversions, the amplified signal is subjected to limiting and then fed to a diode detector (Fig. 3, *a*), which is the most convenient way of obtaining a mean value of d.c. voltage proportional to the number of pulses per unit of time. In the diode detector, each incoming pulse produces a certain charge on the capacitor C_1, which, flowing through a high resistance R_1, causes a voltage U_d on the capacitor which depends on the rate of the pulse or, what is the same thing, on the frequency of the rectangular voltage fed to the integrator.

The relationship $U_d = \varphi(f)$ between the voltage of the diode detector and the frequency of the input pulses can be made linear in the corresponding frequency band by the suitable choice of circuit parameters.

To ensure only slight fluctuation of the output voltage, the time constant $T_c = C_2R_1$ must be considerably greater than the period of the pulses to be measured. In this case, putting $C_2 \gg C_1$ the output voltage is found by the formula

$$U_d = E\,\frac{fC_1R_1}{1+fC_1R_1}, \tag{2-1}$$

where E is the amplitude of the input voltage.

The characteristic $U_d = \varphi(f)$ (Fig.3, *b*) is non-linear owing to the presence in the denominator of the term fC_1R_1. To make it approximately linear, it is necessary to reduce the value of $T_d = C_1R_1$ so that $fC_1R_1 \ll 1$. But in that case the output voltage U_d also decreases correspondingly, which leads to a reduction in the sensitivity of the system. It is therefore necessary to make the value of T_d such that the maximum possible sensitivity is ensured for specified values of deviation from linearity in the characteristic $U_d = \varphi(f)$.

This selection can be made as follows. When frequency modulated rectangular oscillations are fed to the diode detector, the operating band of the characteristic is limited to $f_0 \pm \nu f_0$, where f_0 is the central carrier frequency and ν the modulation factor.

In this case the output voltage U_{do} of the diode detector can be regarded as zero at the frequency f_0. With modulation of the output voltage, there is a difference between the voltage U_d at the frequency f and the voltage U_{do} at the frequency f_0 (Fig. 3, *b*).

The values of the voltages at the extreme frequencies f_{high} and f_{low}

$$f_{high} = (1 + \nu) f_0;$$
$$f_{low} = (1 - \nu) f_0,$$

are expressed as

$$U_{d.high} = E\left[\frac{(1+\nu)f_0 T_d}{1+(1+\nu)f_0 T_d} - \frac{f_0 T_d}{1+f_0 T_d}\right];$$
$$U_{d.low} = E\left[\frac{(1-\nu)f_0 T_d}{1+(1-\nu)f_0 T_d} - \frac{f_0 T_d}{1+f_0 T_d}\right]. \quad (2\text{-}2)$$

where T_d is $C_1 R_1$.

In order, therefore, to determine the degree of nonlinearity in the operating characteristic of the detector, it is necessary to calculate the values of ΔU_{high} and ΔU_{low}, defined as the differences between a tangent to the characteristic $\varphi(f)$ at the point f_0 and the characteristic itself at the points f_{high} and f_{low}.

The tangent at the point f_0 has the following angle coefficient

$$\sigma = \tan\delta = \frac{dU_{f0}}{df} = E\frac{T_d}{(1+f_0 T_d)^2}, \quad (2\text{-}3)$$

and intercepts a voltage U_{dk} on the ordinate axis

$$U_{dk} = E\frac{(f_0 T_d)^2}{(1+f_0 T_d)}. \qquad (2\text{-}4)$$

When considered with (2-2) and (2-3), the values ΔU_{high} and ΔU_{low} are obtained in the form

$$\Delta U_{high} = U_{d.high} - \sigma\nu f_0 = -E\frac{(T_d f_0 \nu)^2}{(1+f_0 T_d)^2\,[1+T_d f_0(1+\nu)]} \qquad (2\text{-}5)$$

and

$$\Delta U_{low} = U_{d.low} - \sigma\nu f_0 = E\frac{(T_d f_0 \nu)^2}{(1+f_0 T_d)^2[1+T_d f_0(1-\nu)]}.$$

The ratios $\Delta U_{high}/U_{d.high}$ and $\Delta U_{low}/U_{d.low}$ are the relative values of the error in the performance of the system due to deviation of the operating characteristic from the linear and they must not exceed a specified value ε depending on the purpose which the system is required to serve:

$$\left|\frac{\Delta U_{high}}{U_{d.high}}\right| = \left|\frac{\Delta U_{low}}{U_{d.low}}\right| = \frac{\nu f_0 T_d}{1+f_0 T_d} \leqslant \varepsilon \qquad (2\text{-}6)$$

or

$$f_0 T_d \leqslant \frac{\varepsilon}{\nu-\varepsilon}.$$

Expression (2-6) allows the required parameters R_1 and C_1 to be selected without difficulty for the required operating band of frequencies $2\nu f_e$ and a specified error ε in operation.

The frequency demodulator circuit without time conversion is very simple, but owing to the nonlinearity of the detector characteristic $U_d = \varphi(f)$ the variation of the output voltage is limited to ± 1.5 to 2 V.

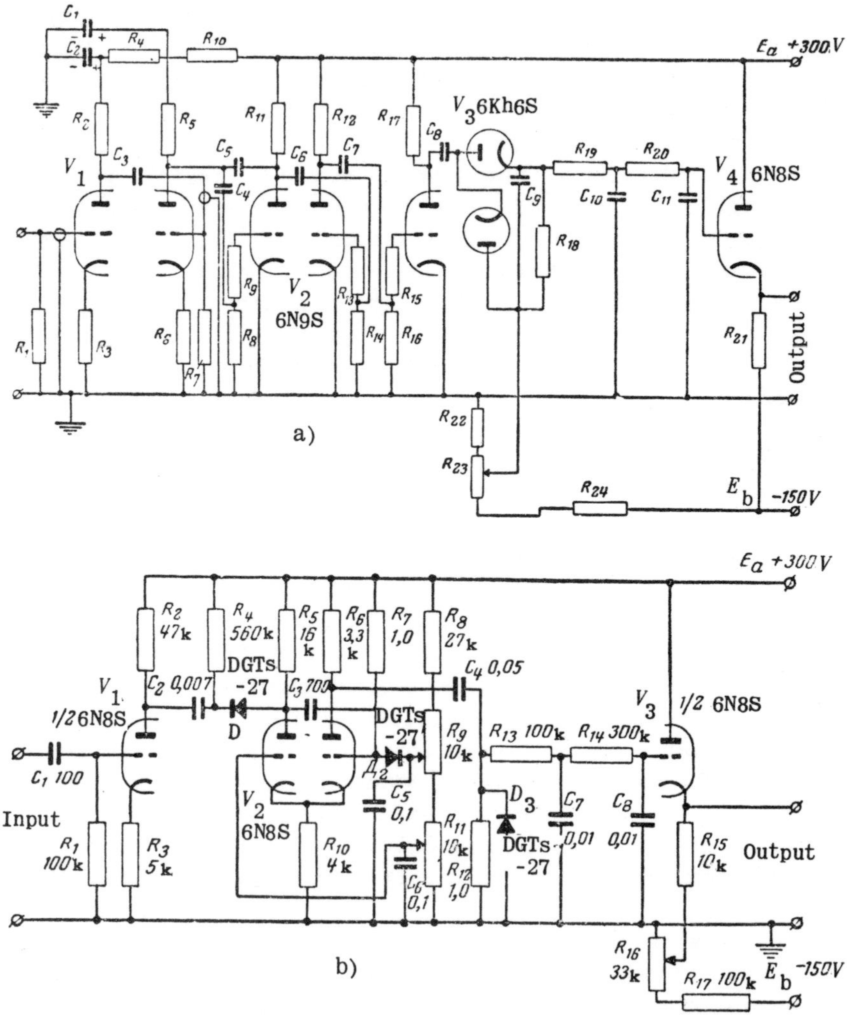

Fig. 4. Circuit diagrams of playback units.

a - with a diode detector:

$R_1 = 220$ k;
$R_2 = R_5 = R_{11} = R_{12} = 360$ k;
$R_3 = R_6 = 1,2$;
$R_4 = R_{10} = 51$ k;
$R_9 = R_{13} = R_{15} = 500$ k;
$R_{17} = 5,1$; $R_{19} = 100$ k;
$R_{20} = 200$ k;
$R_{21} = R_4 = 82$ k;
$R_{22} = 33$ k; $R_{23} = 22$ k;
$R_7 = R_8 = R_{14} = R_{16} = 1,0$; $R_{18} = 510$ k;

$C_1 = C_2 = 20.0 \times 300$ pF
$C_3 = C_6 = C_7 = C_9 = 0.05$ μF
$C_4 = 0.02$ μF
$C_5 = 200$ pF
$C_8 = 82$ pF
$C_{10} = C_{11} = 0,01$ μF

b - with time conversion.

(k = k Ω)

A circuit diagram for a playback unit with this type of frequency demodulator is shown in Fig.4, *a*. The signal from the playback head is amplified by amplification stages (the valve V_1) and then fed to a limiter (the valve V_2). The resulting rectangular pulses are transmitted through a diode detector V_3 which produces a d.c. voltage proportional to the rate of the pulses at its input. The diode detector produces a voltage which is identical to the input voltage of the recording unit, except that it is displaced in time by an amount τ. This is transmitted through a low pass filter to the output element, which is a cathode follower (the valve V_4), and thence to the output of the playback unit. The setting of the output of the playback unit to zero is performed with the aid of the potentiometer R_{23}, whilst the magnitude of the output voltage is set by selection of the resistance R_{18} or the capacitance C_8.

On the other hand, for a playback unit with a frequency demodulator using the time conversion system, the signal from the playback head is amplified and limited in a similar way to that described above (by V_1 and V_2 in Fig.4, *a*), but it is then fed to a differentiating circuit (C_1,R_1 in Fig 4, *b*). The resulting differentiated pulses serve to trigger a monostable multivibrator (the valve V_2) which produces a train of pulses which are of identical amplitude and width, but the distance between them is different depending on the value of the carrier frequency. After these pulses have been transmitted through a rectifier and a low pass filter (D_3, R_{12}, R_{13}, C_7, R_{14}, C_8), a voltage is produced which is identical to the input voltage of the recording unit. The output element V_3 ensures a low output impedance for the playback unit. The purpose of the potentiometer R_{16} is to set the output of the controlled-delay device to zero.

2.4 The Drive

To adjust the delay, it is necessary to be able to alter the speed of the tape and maintain it with the greatest possible accuracy, when the parameters which determine the operating conditions of the device vary (e.g. the load, supply voltage etc.). Moreover, the tape speed must be altered by varying the voltage in order to ensure that this speed matches the respective parameters of the control system or model in which the device is used, e.g., with the speed of billet in a rolling mill.

Two systems are commonly used, but neither meets these requirements satisfactorily. In the first System (Fig.5), motor speed is controlled by varying the armature supply voltage of the motor by an electronic amplifier. The desired speed is maintained by an automatic control system which comprises an electronic amplifier and a tachometer generator which is connected to the motor shaft. This system allows different motor speeds to be set and maintained with a high degree of accuracy. The drawback is that the supply of the armature circuit from an amplifier and this requires that the output power of the amplifier must be comparatively large, which rather complicates the system.

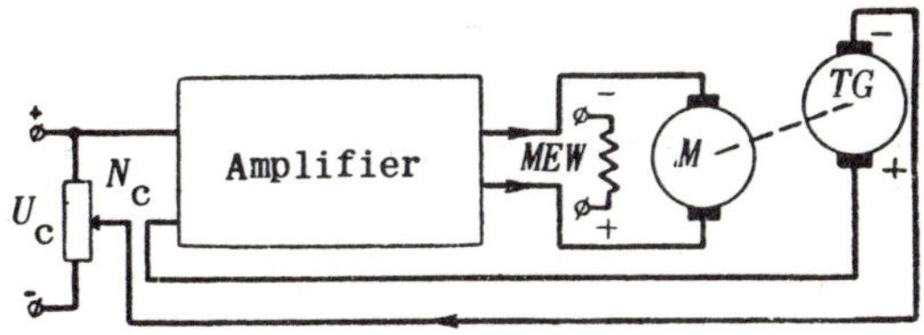

Fig. 5. Circuit diagram for the speed control of the motor by varying the motor armature voltage.

MEW – Motor excitation winding; u_c – control voltage; n_c – control potentiometer.

In the other system (Fig.6) the speed of the motor is controlled by varying its excitation current. But the speed can only be varied over a limited range and the non-linear relationship between excitation current and rotational speed makes it difficult to match the speed exactly with the parameters of the respective analogue or control system.

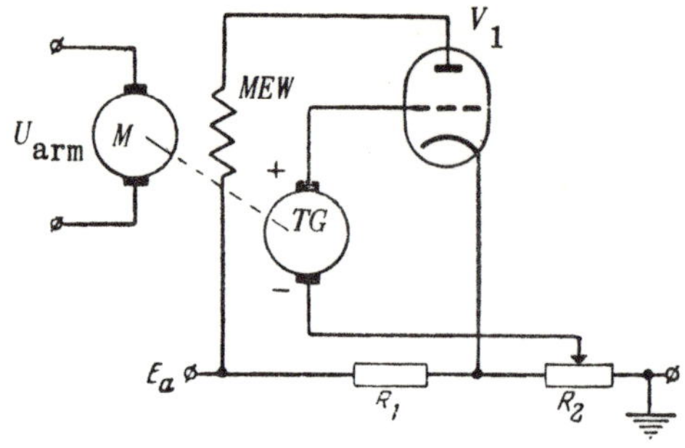

Fig. 6. Circuit diagram for the speed control of the motor by varying the motor excitation current

U_{arm} – armature voltage: *MEW* – motor excitation winding; V_1 – valve; *TG* – tacho-generator

The shortcomings of both these systems have been eliminated in a motor speed-control system which has been specially developed for controlled delay devices (Fig. 7). Here the motor armature voltage is varied by a barretter in the motor armature circuit and a low-power amplifier which supplies the motor excitation winding. The system exhibits a linear relationship between the motor speed and the control voltage which allows the delay to be matched with the parameters of the other devices without difficulty.

The system works as follows. On variation of the control voltage U_c the current through the excitation winding is changed which with a constant (unchanged) torque on the motor shaft leads to a change in the armature current. Since a barretter is connected in series with the motor armature, even a small change in the armature current leads to a considerable change in the voltage drop across the barretter and as a result the armature voltage of the motor and its rotational speed also change. If U_c increases, the excitation current of the motor is also increased, whilst the armature current is reduced, which reduces the voltage drop

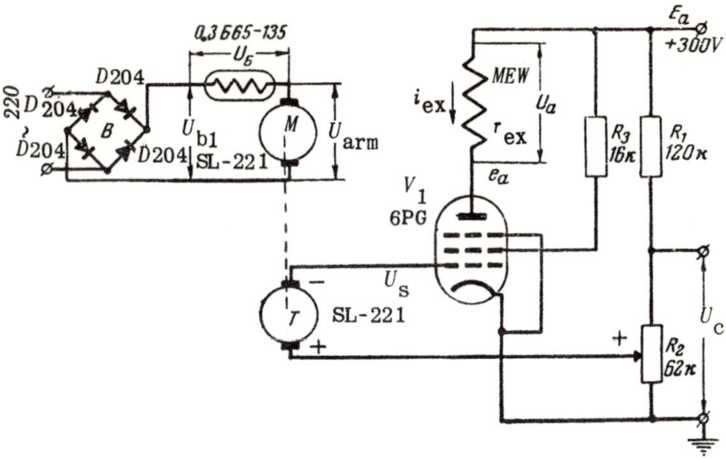

Fig. 7. Circuit diagram for the speed control of the motor by a barretter

r_{ex} – excitation resistance; U_{arm} – armature voltage.

across the barretter and increases the armature voltage with the result that the rotational speed of the motor is also increased. But since the difference between the control voltage and the voltage of the tacho-generator in the feedback circuit is fed to the amplifier, the speed of the motor will vary until the value of the bias voltage on the grid of the amplifier valve attains a value approximately equal to its original value. Thus the change in control voltage leads to a proportional change in the motor armature voltage and the relationship between the speed of the motor and the control voltage in the system is linear. If the load torque on the motor shaft is constant in value, the excitation current remains practically constant at different specified motor speeds. The set speed is not altered by variations of load torque or mains voltage in view of the presence of rate feedback.

2.5 Technical Characteristics and Construction

Controlled-delay devices using magnetic recording allow time delays of 0.1 sec to 20 min to be obtained, but this time can be considerably increased if a special tape-tensioning attachment and drive are used. The relative error in the reproduction of delayed signals is about 1.5 per cent in delay devices using the standard tape-tensioning attachment. This error can be reduced by precision production of the drive roll of the tape-tensioning device. The band width of the transmitted frequencies is 0 to 20 c/s. This can be increased to 200 c/s on increasing the carrier frequency to 5000 c/s.

Fig. 8. A controlled-delay device using magnetic recording.

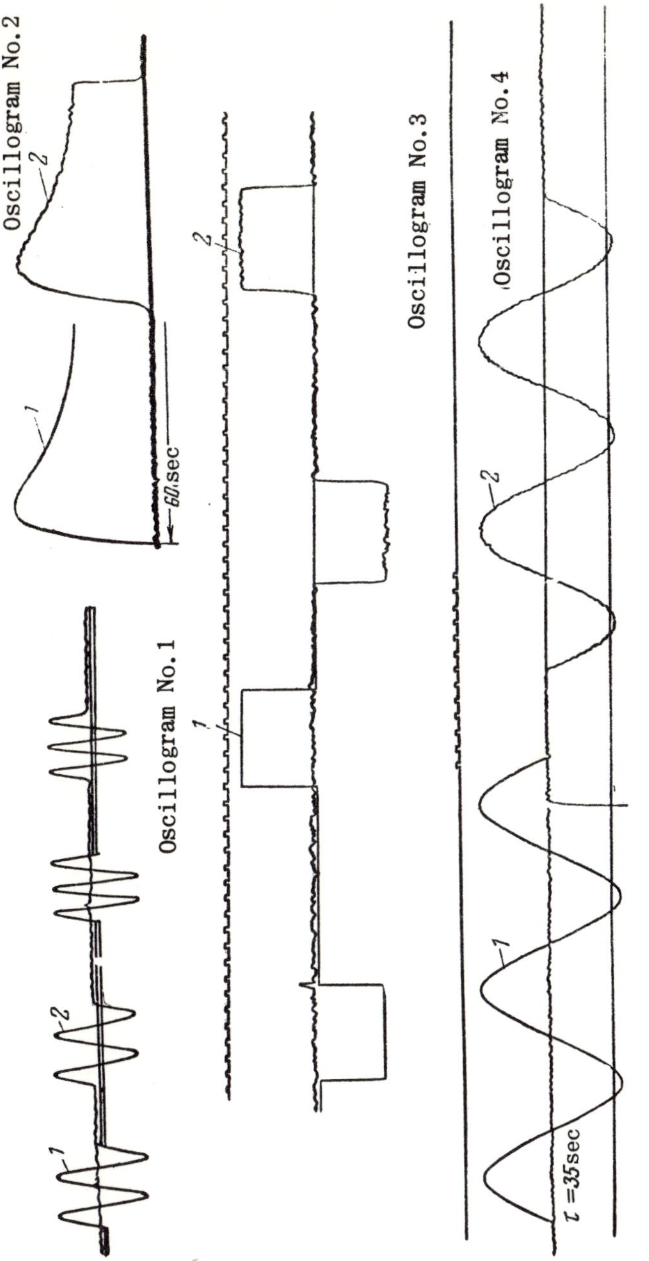

Fig. 9. Oscillograms of typical signals.
1 - original signal; 2 - delayed signal

Figure 8 shows a photograph of a controlled-delay device using magnetic recording which has been developed and produced at the Institute of Automation and Remote Control of the U.S.S.R. Academy of Sciences. Figure 9 shows oscillograms of the typical input and output signals of this delay device. The recording, playback and drive units of this particular device are made in accordance with the circuits of Figs. 2, *a*, 4, *a* and 7 respectively.

As regards practical design, delay devices are usually made in four separate compartments which contain the recording unit, playback unit, the electronic part of the drive unit, and the tape-tensioning device with motor and tacho-generator of the drive unit respectively. These sections are mounted independently of each other. This arrangement facilitates the production of the delay device and its assembly.

The continuous reel to which reference has been made, can be in the form of rollers through which an endless loop of magnetic tape passes for comparatively short time lags, or in the form of a disk with a small cylinder in the middle part (see Fig.8) on which the tape is wound.

The tape is fed to a spool in the form of a disk with a short cylinder, and it leaves the spool from the back of the cylinder. When originally wound, the roll of tape must not be pressed tightly to the cylinder. This is prevented by inserting a packing between the tape and the cylinder which is afterwards removed.

The tape is driven by a capstan, connected to the motor, to which the tape is pressed by a rubber-coated roller. The spool has no drive, but to ensure that the tension of the tape is good, the spool is slightly retarded.

To ease the tape as it leaves the back of the spool, the disk and cylinder assembly is set at an angle to the plane of the magnetic heads so that the tape cannot come into contact with the disk on leaving the reel.

CHAPTER III

CONTROLLED-DELAY DEVICES USING STORAGE CAPACITORS

3.1 Design of controlled-delay Devices using Storage Capacitors and Methods of improving the Accuracy of Reproduction

There are several known ways of constructing capacitor controlled-delay devices. They are all based on the ability of a capacitor to store the charges applied to its plates. The main part of such delay devices is the storage unit which is a large bank of capacitors.

Discrete values of the input signal are stored in this unit and sampled from it after a specified time lag τ.

Such delay devices fall into two categories: *(a)* delay devices with sequentially switched capacitors, and *(b)* delay devices with progressively switched capacitors.

In the first system each capacitor stores a single constant value of the voltage for the duration of the time lag τ and the order of the capacitors is changed in each cycle of operation of the delay device.

In the other system all the voltage values pass progressively through each capacitor and the order of the capacitors remains unchanged.

(a) Controlled-delay devices with sequentially switched capacitors

The storage unit of this type of delay device is a row of capacitors which are electrically connected to a commutator which has two slip contacts for recording and sampling (Fig.10,*a*). Brush No.1 of the commutator rotates in the direction indicated by the arrow and connects the capacitors one after the other to the input voltage, thereby storing discrete values of this voltage during the time the brush with the commutator segment is in contact with the capacitors C_1, ...C_n.

In travelling along the segments of the commutator one after the other, brush No.2 samples the input signal level stored in the capacitor.

The time lag τ is equal to the period of time taken for the two brushes to pass the same segment. It therefore depends on the speed Ω of the brushes and the angle α_0 between them:

$$\tau = \frac{\alpha_0}{\Omega}. \tag{3-1}$$

The shape of input voltage curve $U_{in} = f(t)$ is shown in Fig.10, *b*, and that for the output voltage from brush No.2 $U_{out} = f(t-\tau)$ is shown in Fig.10, *c*. It will be seen from this diagram that the output voltage is a stepped function with a delayed argument.

An instantaneous value of the input signal is thereby stored on each capacitor C_1, ..., C_n at instants Δt apart. The particular interval Δt, which defines the division of the continuous function of the input signal into the series of "steps" which are stored by the capacitors C_1, ..., C_n, depends on the speed of the commutator brushes and the angle $\Delta\alpha$ of each segment:

$$\Delta t = \frac{\Delta\alpha}{\Omega}. \tag{3-2}$$

In its turn the angle $\Delta\alpha$ depends on the number of storage capacitors N and the subtended angle α of all the segments of the commutating assembly:

$$\Delta\alpha = \frac{\alpha}{N}. \tag{3-3}$$

The time lag is thus

$$\tau = \frac{\alpha_0}{\alpha} N\Delta t. \qquad (3\text{-}4)$$

It follows from equation (3-4) that the time lag is directly proportional to the number of storage capacitors and the period of time Δt between charging adjacent capacitors. Since Δt is a finite quantity, a stepped voltage curve is produced at the output of the sampling unit (Fig.10, *c*) with the result that error can occur in the amplitude and time lag of the output voltage. The time lag error obviously cannot be greater than the extent of Δt. The amplitude error is the error in the approximation of the continuous function of the input signal by a stepped function with step intervals Δt.

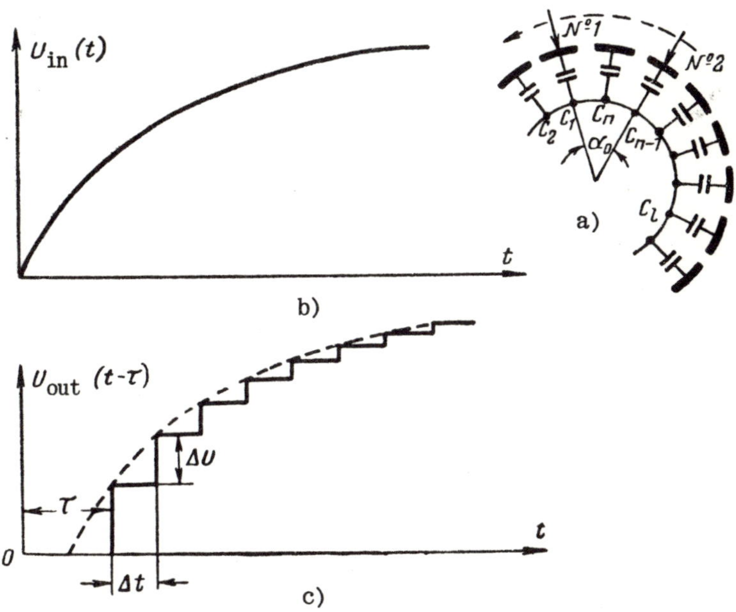

Fig. 10. The storage unit of a controlled-delay device and reproduction of a typical input signal

a - storage unit; *b* - original input signal; *c* - output signal after the sampling unit.

This error depends on the nature of the variation of the input function and the number of storage capacitors. The larger the number of capacitors to reproduce a particular section of the input signal curve, the smaller is the error. But a limit is naturally imposed on the number of capacitors by the ultimate size of the delay device and the difficulties in making the commutating assembly.

It is therefore desirable to have an estimate of this error for a given number of storage capacitors N, a specified frequency band ω of the input signal and the required range of time delays τ.

This error in approximation can be estimated by the expression

$$\Delta U = \left(\frac{dU}{dt}\right)\Delta t, \tag{3-5}$$

where ΔU is the absolute error, and dU/dt the rate of variation of the curve $U(t)$ in the period Δt.

For a sinusoidal input signal, the maximum value of dU/dt is given by the expression

$$\left(\frac{dU}{dt}\right)_{\max} = U\omega. \tag{3-6}$$

The maximum value of the relative error in the approximation is found by the simultaneous solution of equations (3-4) - (3-6):

$$(\delta U)_{\max} = \left(\frac{\Delta U}{U}\right)_{\max} = \xi \frac{\omega\tau}{N}\, 100\%, \tag{3-7}$$

where $\xi = \frac{\alpha}{\alpha_0}$ depends on the construction and parameters of the commutating assembly.

It follows from (3-7) that the error in question is proportional to the magnitude of $\omega\tau$ and inversely proportional to N.

Figure 11 shows curves which have been plotted from formula (3-7) which indicate that it is necessary to select the frequency ω of the input signal and the controlled delay time τ so as to ensure the desired accuracy in reproducing the input signal. It is supposed that the values of ξ and N have already been selected in the design procedure and that the coefficient M has therefore been defined:

$$M = \frac{\xi}{N}. \tag{3-8}$$

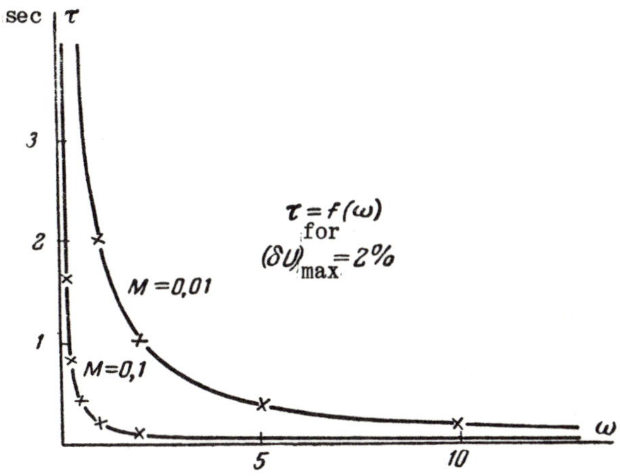

Fig. 11. The time lag as a function of the frequency of the input signal for a 2 per cent error in reproduction.

$M = \xi/N$; $\xi = a/a_0$ (see equation (3-3) and Fig. 10, a); N – the number of storage capacitors.

In Fig. 11 the permissible region of values of ωt lies below the corresponding curve $\tau = f(\omega)$ plotted for several values of M. Thus, for example, a sinusoidal input signal with a frequency $\omega = 5$ cannot be reproduced at the output of a controlled-delay device with less than 2 per cent error (usually considered permissible) for a specified time delay of 1 sec, since the point of intersection for the corresponding values of ω and τ lies above the curve of permissible values.

Using formula (3-7), it is possible to estimate the error in reproducing the input signal for different numbers of storage capacitors N and different values of the quantity $\omega\tau$.

To keep this error within the permissible limits, the controlled-delay device must usually have a large number of storage capacitors, the range of variation of the quantity $\omega\tau$ being comparatively small.

Calculations show that the requisite 2 per cent error in signal reproduction is attained at $\omega\tau = 0.2\pi$ and $N \doteq 100$. For example, an input signal at a frequency of 1 c/s, can only be reproduced with a 2 per cent error if the controlled-time lag is maintained within the range 0 to 0.1 sec.

In order to widen the frequency band of the input signal and increase the delay without increasing the error in reproduction, it is necessary to improve the shape of the output voltage curve by smoothing out the steps. Such smoothing as is achieved by connecting a filter at the output of the delay device leads to distortion of the amplitude phase characteristic of the delay device and reduces the permissible quantity $\omega\tau$.

Hence the method of a piecewise-linear approximation of the stepped output voltage curve was applied.

In this method a linearly varying increment is added to each voltage step. This addition connects the corners of adjacent steps of the curve (see the broken line in Fig. 10, *c*). The resulting curve for the output voltage reproduces the input signal much more closely.

The error in this piecewise-linear approximation can be estimated by the expression [2]:

$$\Delta U = \frac{1}{8}\left(\frac{d^2U}{dt^2}\right)\Delta t^2, \tag{3-9}$$

where ΔU is the absolute error; and
d^2U/dt^2 – the second derivative of the input signal function $U(t)$.

For a sinusoidal input signal, the maximum relative error is

$$(\delta U)_{max} = \left(\frac{\Delta U}{U}\right)_{max} = \frac{1}{8}\left(\xi\frac{\omega\tau}{N}\right)^2 100\%. \qquad (3\text{-}10)$$

Calculations performed by this formula show that a 2 per cent error in the input signal reproduction is achieved if $\omega\tau = 2\pi$. The piecewise-linear approximation thus widens the range of values of $\omega\tau$ by a factor of 10 as compared with the ordinary stepped approximation (other factors being equal).

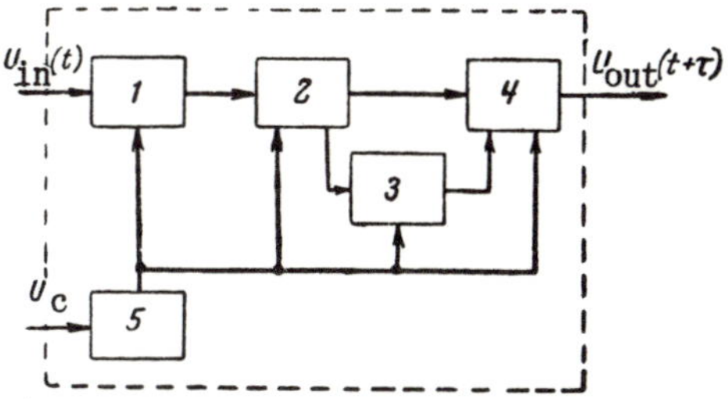

Fig. 12. Block diagram of a controlled-delay device using sequentially switched storage capacitors

1 – input element and recording unit; *2* – storage unit; *3* – unit for the piecewise-linear approximation of the output voltage curve; *4* – output element and the sampling (reproduction) unit; *5* – unit for controlling the delay τ, (u_c – the control voltage).

A block diagram of a controlled-delay device incorporating sequential commutation of the storage capacitors and piecewise-linear approximation of the stepped voltage curve is given in Fig. 12.

A cathode follower or an operational d.c. amplifier is

usually used as the input element, which then has a large input impedance and a small output impedance. This ensures rapid charging of the storage capacitors and a low consumption of current from the recording source.

The recording unit, like the sampling unit, is a commutator with two slip contacts, for which a step selector can be used. In view of the possibility of the capacitors discharging through the insulation resistance of the slip contacts, the insulating material, the moving and static contacts are fastened on, must be of high quality (usually organic glass).

It is best to use storage capacitors with a polystyrol or styroflex dielectric (types KPG or PGS).

The sampling unit must possess a very large input impedance to prevent the discharge of a storage capacitor in the sampling process; its output impedance must be as low as possible to connect easily the controlled-delay device with the rest of the apparatus.

Figure 13, *a* illustrates the system in which an operational d.c. amplifier is used for the output of the sampling unit. In this diagram the storage capacitors C_1, ..., C_n are connected in turn in the feedback circuit of the operational d.c. amplifier parallel to the capacitor C_0. The capacitance C_0 is much less than that of the storage capacitors. This capacitor is needed to maintain the sampled voltage when the feedback circuit is otherwise broken at the instant of switching. The discharge of the capacitor C_0 in the interval between sampling cycles depends on the leakage resistance R_{leak}, and is quite small since the discharge time constant is large:

$$T = R_{leak}C_0(1 + k_a), \tag{3-11}$$

where k_a is the gain of the d.c. amplifier A (usually $\geqslant$ 50,000).

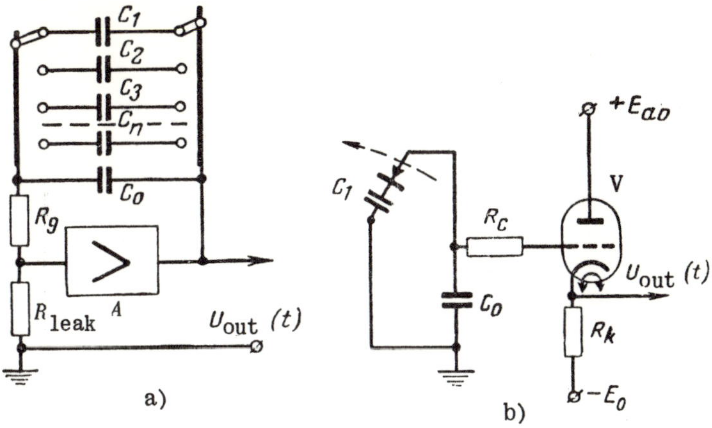

Fig. 13. Principle circuits for sampling the voltage from the storage unit

a - using an operational amplifier; *b* - using a cathode follower.

R_g - a resistance to prevent discharge of C_0; R_{leak} - leakage resistance; R_k - cathode circuit resistance; R_c - a large resistance to limit the grid current in the grid circuit of the cathode follower; V - valve; (explanations in text).

A resistance R_g can be used in the circuit of the capacitor C to reduce the discharge of the capacitor C_0 at the instant of commutation. In some controlled-delay devices [5], a cathode follower is used instead of the operational d.c. amplifier (Fig. 13, *b*). This also has a high input impedance and low output impedance.

In this system, the storage capacitors C_i are connected in turn to the input of the cathode follower in parallel with the capacitor C_0 which performs the same functions as the capacitor C_0 in Fig.13, *a*.

To limit the grid current, a resistance R_c of rather large magnitude (of the order of 10 M Ω) is introduced in the grid circuit of the cathode follower, but the choice of the valve type and its operating conditions are selected with a view to ensuring conditions which are "close to the 'electrometric' state" [2].

These sampling units lead to error in signal reproduction owing to the finite magnitude of the capacitance C_0.

The amount of error depends on the relationship between the capacitances C_i of the storage capacitors and C_0, the length of the period Δt and the nature of the variation of the input signal voltage:

$$\Delta U_s = \frac{C_0}{C_i + C_0}\left(\frac{dU}{dt}\right)\Delta t. \qquad (3\text{-}12)$$

For the system of Fig.13, *a*, simultaneous solution of equations (3-4), (3-6) and (3-12) results in the following expression for the maximum relative error:

$$(\delta U_s)_{\max} = \frac{C_0}{C_i + C_0}\left(\xi \frac{\omega\tau}{N}\right) 100\%. \qquad (3\text{-}13)$$

Calculations which have been performed using formula (3-13) show that the capacitance C_0 must be of a lower order than C_i to ensure satisfactory accuracy in operation.

The unit for the piecewise-linear approximation of the output voltage curve incorporates a device for obtaining the difference between two adjacent stored values of the step function and a device for producing corresponding linearly-varying voltages. The difference in the values of the two voltages is usually obtained by operational d.c. amplifiers, one of which is an inverter and the other an adder.

Figure 14, *a* illustrates a different method of obtaining the difference in the values of the stored voltages. Here the storage capacitor C in one of the cycles when the contact K_1 of the controlling unit is closed, is charged to a magnitude of voltage equal to the corner of the step of the function at the output of the sampling unit. In the next cycle, to which a new corner of the step function corresponds, the contact K_1 is opened and as a result the difference in values of the voltages of adjacent steps is produced at the output.

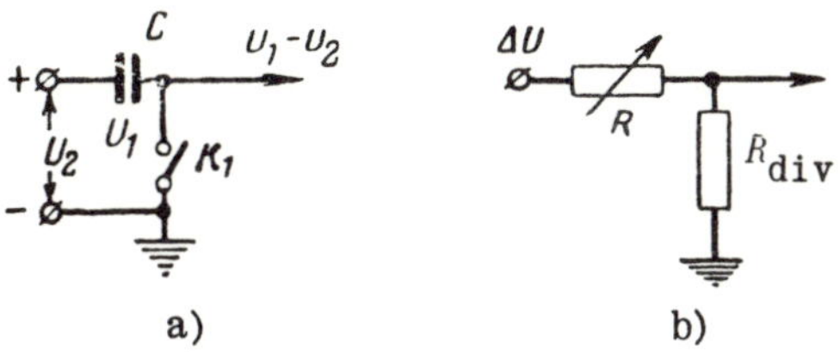

Fig. 14. Circuit for adding a linearly-varying increment to the output voltage:

a – storage circuit; *b* – voltage divider.

K_1 – contact of control unit; *C* – storage capacitance; R_{div} – voltage divider.

The device for forming the linearly varying increments of the approximating voltage can also be made in several different ways. One way is to use an operational d.c. amplifier as an integrator.

This integrator receives rectangular voltages ΔU from the output of the device for obtaining the difference in voltages at adjacent steps of the function. The output voltage e_{out} of the integrator is defined by the expression

$$e_{out} = \frac{1}{RC} \int_0^{\Delta t} \Delta U \, dt = \frac{\Delta t}{RC} \Delta U, \tag{3-14}$$

where RC is the time constant of the integrator.

If $RC = \Delta t$, the voltage of the linear increment reaches the necessary value ΔU after a period of time Δt, and the output voltage therefore varies along a straight line joining the corners of adjacent steps in summation with the stepped curve.

Since the value of Δt alters when the delay τ is changed, the time constant RC must be automatically matched with the value of τ [6].

Another type of unit for forming the linearly-varying increments is based on saw-tooth modulation of rectangular voltage pulses ΔU.

In Fig.14, *b* the voltage pulses ΔU are fed to a voltage divider R_{div} which consists of series-connected resistors, the upper-one of which varies linearly with time during the period Δt. Since the value of Δt alters when τ is changed, the rate of variation of the resistance R of the modulator must be automatically matched with the value of τ. The method of matching will be considered in more detail below.

The task of the control unit of the controlled-delay device is to set the time lag τ in accordance with the magnitude of the external control voltage U_0.

In the system of Fig.12 the delay τ depends on the rotational speed Ω of the brushes of the commutating assembly and the control unit is a servo system which ensures that the speed of the commutator brushes is proportional to the control voltage U_c. This servo system will be considered in more detail in the last section.

The output unit of the controlled-delay device serves to match the delay device with the input to the other apparatus, for additional smoothing of the output voltage curve and correcting the required gain of the delay device (the gain must be equal to unity). Another purpose of the output unit is to summate the voltages of the step function with the linear increment of the approximation voltage.

A four-pole network with a large input and small output impedance can be used as the output unit. This is usually an operational d.c. adding amplifier with a small capacitance for final "smoothing" of the output voltage curve.

The capacitance is connected in the feedback circuit of the amplifer in parallel with the variable resistance in order to correct the gain.

(b) Controlled-delay devices with sequential transmission of the input signal along "a chain of storage cells"

The storage unit of this type of controlled-delay device [8] is a "chain of storage cells" which are separated from each other by contacts. All the even contacts of the set are normally closed and all the odd are normally open (Fig. 15).

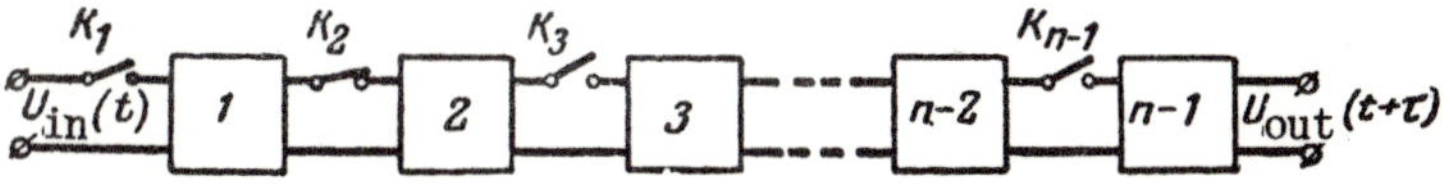

Fig. 15. Block diagram of the storage unit of a controlled-delay device with sequential transmission of the input signal along a chain of storage cells.

The principle of the device is that the commutation of the contacts forces the input voltage to spread along a row of series-connected storage cells and charge a new cell in each cycle.

By way of illustration, suppose that all the storage capacitors of all the cells are uncharged and that the delay device receives an input signal voltage of constant magnitude. In the first cycle all the odd contacts are then closed and all the even contacts are open. This leads to the charging of the storage capacitor of the first cell to the voltage of the input signal. The capacitors of the rest of the odd cells will not be charged since the preceding storage cells have not been charged. In the next cycle all the odd contact sets open and all the even ones close. In this case, the input signal voltage stored on the capacitor of cell No.1 is transmitted to cell No.2 where it charges the storage capacitor there. The capacitors of the rest of the even cells again will not be charged since the preceding storage cells have not been charged.

A new cycle of operation again leads to the closing of all the contact sets of the even cells. The input signal voltage now proceeds to cell No.3 where it charges the storage capacitor. The input voltage in this case also proceeds to the storage capacitor of cell No.1 as well, but this capacitor has already been charged to the same amount and now the capacitance of cell No.1 is only charged to the new magnitude of the input signal voltage (if this voltage has changed during the first two cycles).

The rest of the cells in the storage unit do not receive voltages in the cycle of operation under consideration.

The input signal voltage continues along the chain of storage cells in this way until it reaches the output of the storage unit after N cycles of the control circuit, the number N here being equal to the number of storage cells in the chain.

The time lag τ depends on the number of storage cells N and the commutation frequency of the contact sets

$$\tau = \frac{N}{f} = 2N\Delta t, \qquad (3\text{-}15)$$

where Δt is the period during which the contacts of the storage unit are closed (open).

In this type of system, the delay is controlled by altering the rate at which the contact sets of the storage unit are switched.

Owing to the finite value of the period Δt in between the charging of numerically adjacent storage capacitors, a stepped and not a continuous output voltage curve is stored and reproduced in this type of delay device. The width of the steps of this curve is conditioned by the rate at which the contact sets are switched. The stepped shape of the output voltage curve leads to an error in amplitude and delay which can be determined by the methods considered previously.

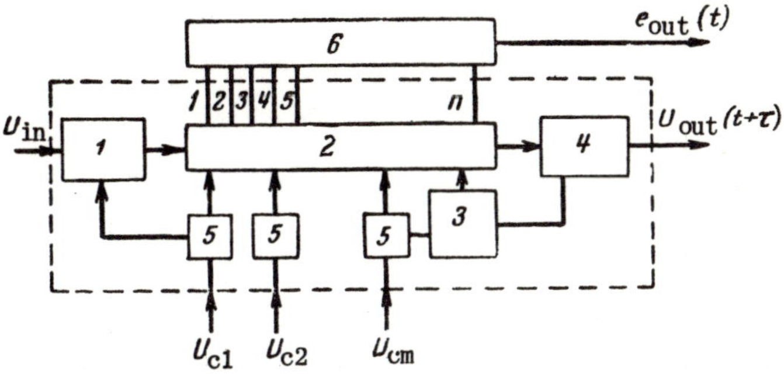

Fig. 16. Block diagram of a controlled-delay device with sequential transmission of the input signal along a chain of storage cells.

1 - input element; *2* - storage unit; *3* - unit for the piecewise-linear approximation of the output voltage curve; *4* - output element; *5* - unit for controlling the delay τ; *6* - adding device required in the synthesis of a specified transfer function.

The introduction of piecewise-linear approximation in delay devices of this type can obviously improve the accuracy with which the input signal is reproduced and widen the range of values of the quantity $\omega\tau$.

Figure 16 shows a block diagram of a controlled-delay device using the principle of forced propagation of the input signal along a chain of storage cells and incorporating piecewise-linear approximation of the stepped output voltage curve.

Like the system with sequential switching (Fig.12),

the input element in this type of system could be an operational d.c. amplifier or a cathode follower.

The piecewise-linear approximation is also obtained by the same methods as before (Fig.14).

The main feature of delay devices with sequential transmission of the input signal is the absence of a special commutator, since the storage cells are switched by means of relay contacts (or electronic switches), which is undoubtedly one of the advantages of this type of controlled-delay device. Another advantage is the absence of a special sampling unit, since the input signal voltage is transmitted along the chain of storage cells by the relay contacts referred to above.

The time lag τ is adjusted by altering the switching rate of the contact sets in the chain of storage cells and the means of control is the circuit of the frequency modulator.

The frequency modulator ensures that the rate of the output voltage pulses is proportional to the magnitude of the control voltage U_c.

The electronic apparatus of a frequency modulator is much simpler than the elements of a velocity feedback control system which includes a motor, a tacho-generator and other electro-mechanical devices.

This is yet another advantage of the type of delay device in question.

A controlled-delay device with sequential transmission of the signal along a chain of storage cells can also be used for the synthesis of a specified transient transfer function with the addition of an adder [12]. For instance, this problem arises in the simulation of a unit from an experimentally-determined transient transfer function. It is common knowledge that the transient transfer function defines the dynamic properties of a unit and represents the response of a system to a step input. It is therefore necessary to connect the elements of the electrical analogue and select their parameters in such a way that the response of the electrical analogue to the step input corresponds to the experimentally-determined curve for transient behaviour. In a few simple cases no difficulties arise in selecting the structural scheme of a model and calculating its parameters. For example, an exponential transient transfer function corresponds to the inertial elements of an electro-model.

In more complicated cases, however, it is a very tedious business selecting these elements and their parameters. A great deal of time is involved and special skill is called for.

Controlled-delay devices are the answer insofar as electro-models of linear elements are concerned. Hence, the output voltage of each storage cell in the chain is fed to an adding device (Fig.16). The unit steps in the input voltage of the delay device are propagated along the chain of storage cells, charging a new cell after each cycle of operation. The unit steps in the input voltage, therefore, appear first at the output of the first storage cell, then at the output of the second and so on. The output voltage of the adding device also varies correspondingly.

The characteristic of the output voltage is a stepped curve.

The variation of the scale coefficients with respect to each input of the adding device results in a change in the shape of its curve for the output voltage $e_{out}(t)$.

The appropriate choice of scale coefficients ensures that the output voltage curve of the adding device is close in shape to the required transient transfer function.

In the general case, controlled-delay devices of the type under consideration (Fig.16) can have several control devices(U_{c1}, U_{c2}, U_{c3}, ..., U_{cm}), each of which controls its own section of the chain of storage cells. This introduces greater flexibility into the shaping of the output voltage curve of the adding device.

The type of capacitor controlled-delay device which has just been considered in reference to the synthesis of a transient transfer function is also being used in automatic control systems as a correcting element.

3.2 Description of the System and individual Units

We will begin our description of capacitor controlled-delay devices with simple circuits having commutating assemblies in which the input signal is sampled without piecewise-linear approximation of the output voltage curve. Such devices are usually used as attachments to electronic analogue apparatus. Their input and output elements are therefore based on operational d.c. amplifiers.

The fundamental circuit of one such device is shown in Fig.17. The circuit includes a number of storage capacitors(N = 50) which are connected to the segments of a step selector which is driven by a two-phase motor.

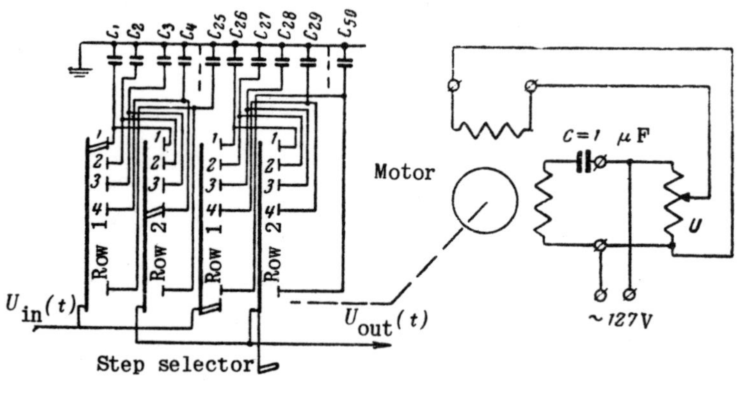

Fig. 17. Circuit diagram of a controlled-delay device using storage capacitors.

The segments of the selector are secured by means of an insulating bush on the gear shaft of the motor and they turn relative to each other at an angle of the order of 30°.

One of these segments is used for recording the input signal voltage on the storage capacitors, but the other, which lags behind it by 30°, is used for sampling the voltage level.

The speed of the motor shaft is changed by varying the voltage in the control winding by an auto-transformer.

A photograph of this delay device is shown in Fig. 18. The step selector, its rotating segments on the gear shaft and the row of storage capacitors are clearly visible.

Fig. 18. Photograph of a simple capacitor controlled-delay device.

1 – two-phase motor with reduction gears; *2* – step-by-step switch; *3* – storage capacitors; *4* – rotating segments of the step-by-step switch, mounted on the shaft of the reduction gear of the motor.

The step-by-step switch can also be driven direct from electro-magnets (without altering the selector) which obtain supply from the frequency modulator. Control can thereby be exercised over the speed of the rotating segments. A change in the control voltage at the input to the frequency modulator alters the frequency of the voltage pulses at its output and therefore alters the frequency of the pulses in the coil circuit of the step-by-step switch.

The circuits of frequency modulators have been described quite fully [3, 4] and this will not be repeated here. (The english-speaking reader is referred to a short book called "Frequency Modulation" by L.B. Arguimbau and R.D. Stuart, Methuen). The modulator circuit of Fig.2 *a* can be used by increasing the time constants of the grid circuits of the multivibrator as appropriate. Instead of the recording head, a relay is set up at the output of the circuit to commutate the electromagnet circuit of the step-by-step switch.

The control of the time lag by varying the d.c. control voltage enables this delay to be matched quite easily with the parameters of an electronic analogue or with those of a control plant, e.g., by means of a tacho-generator.

We will now consider the construction and circuit of the controlled-delay device of Fig.19 which features sequentially-switched storage capacitors and a unit for piecewise-linear approximation of the output voltage curve.

On rotating the rotor of the commutator, the storage capacitors $C_1, \ldots, C_{30}$ are practically instantaneously charged via the contact K-1 and the input d.c. operational amplifier A_1 to the instantaneous value of the input voltage. The sampling contacts K-2 and K-3 of the commutator connect the cathode follower (the valve V_1) to two adjacent storage capacitors. The output voltages of the two cathode followers are shifted relative to each other by the time Δt taken for the sampling contact to pass one segment of the commutator (see (3-2)), and relative to the input signal by the extent of the time lag τ and $\tau - \Delta t$ respectively.

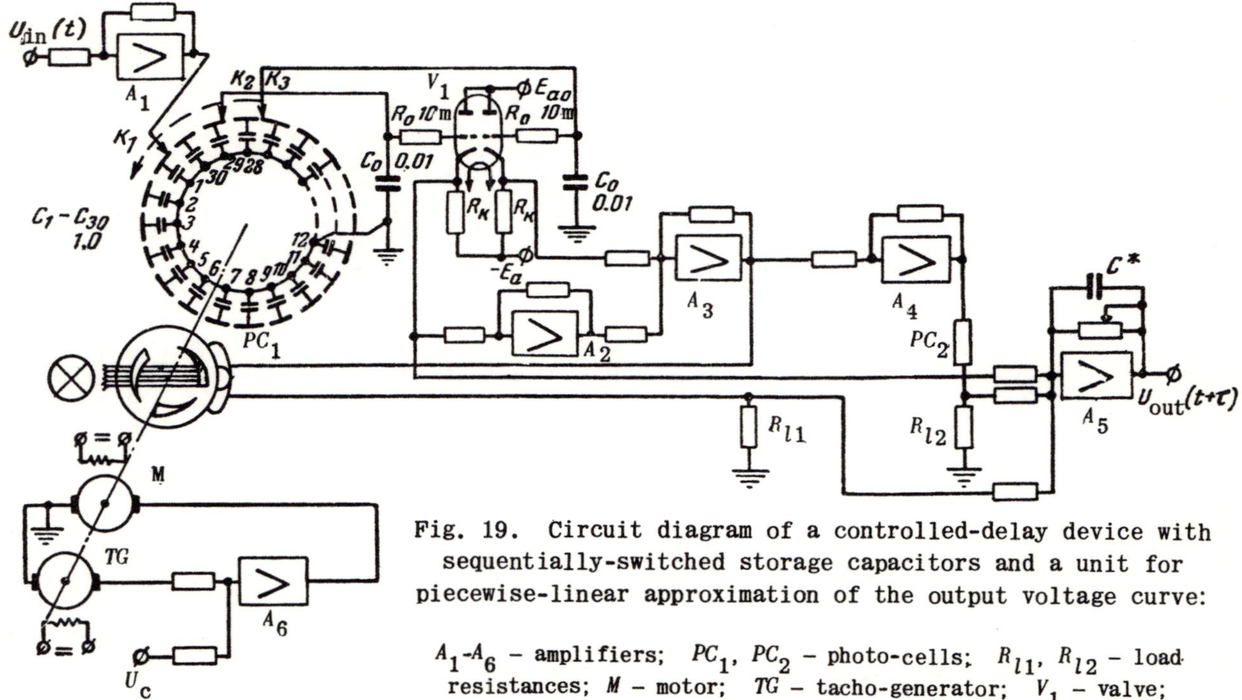

Fig. 19. Circuit diagram of a controlled-delay device with sequentially-switched storage capacitors and a unit for piecewise-linear approximation of the output voltage curve:

A_1-A_6 - amplifiers; PC_1, PC_2 - photo-cells; R_{l1}, R_{l2} - load resistances; M - motor; TG - tacho-generator; V_1 - valve; U_c - control voltage. (explanation in text)

These output voltages are fed to the piecewise-linear approximation unit where the difference between them is found by two d.c. operational amplifiers, one of which (A_2) is used as an inverter, and the other (A_3) as an adder for the input voltages.

The required saw-tooth voltage of the linear increment is obtained in this circuit by a photo-electric modulator which consists of a light source and a photo-cell PC_1, between which a disk is rotated with saw-tooth slots which is coupled to the shaft of the rotor of the commutator. On rotating this rotor, the photo-cell PC_1 is periodically illuminated by a linearly-increasing flow of light which passes through the slots of the disk and causes "saw-tooth" variation in the illumination of the photo-cell and a corresponding change in the amount of its resistance.

The photo-cell PC_1 and load resistance R_{l1} are connected at the output of the operational amplifier A_3. A change in the photo-cell PC_1 leads to a linear variation of the current in the circuit of the load resistance R_{l1}, which ensures the formation of the corresponding saw-tooth voltage for the linear increment at the output of the circuit. The dark current of the photo-cell PC_1 is compensated by the d.c. operational amplifier A_4 operating as an inverter, the photo-cell PC_2* and the load resistance R_{l2}.

The voltage from the output of the resistance R_{l2}, which is proportional to the dark current of PC_2, is fed to one of the two inputs of the output operational amplifier A_5; the other input of this amplifier receives a voltage from the load resistance R_{l1} which is of opposite sign. Compensation of the dark current of the photo-cell PC_1 is thereby achieved. The linear increment voltage from the output of R_{l1} is added in the amplifier A_5 to the output voltage of the step function of the cathode follower V_1, which forms the piecewise linear shape of the output voltage of the delay device (Fig 20)

* The photo-cell PC_2 is of the same type as PC_1 (actually, an FC – Al type soviet photo-cell was used)

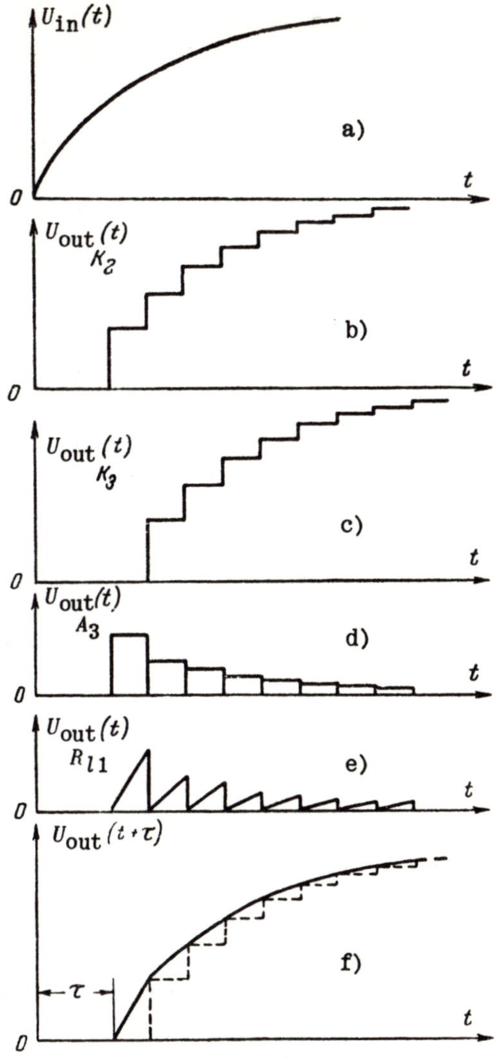

Fig. 20. Formation of the piecewise-linear function at the output of the approximation unit.

Figure 20 illustrates the formation of the piecewise-linear function at the output of the approximation unit and the piecewise-linear shaping of the output voltage of the delay device. The shape of the curve $U_{in}(t)$ for the input voltage of the delay device is shown in Fig.20, *a*. The voltages at the outputs of the cathode followers of the sampling units are illustrated in Fig.20, *b* and *c*. The subtraction of these output voltages by the amplifiers A_2 and A_3 results in the formation of a step function $U_{out.A_3}(t)$ at the output of A_3 (Fig.20, *d*). The saw-tooth modulation of each rectangle of this voltage by the photo-cell PC_1 and the disk with saw-tooth slots leads to the formation of the necessary linear voltage increment at the output of the load resistance R_{l1} (Fig.20, *e*). Finally, Fig.20 *f* illustrates the shape of the output voltage curve of the delay device.

The unit for controlling the time lag τ is a speed servo system (Fig.19), which consists of a d.c. motor M, a speed measurer (the tacho-generator TG), and a d.c. amplifier A_6, which is similar to the other amplifiers A_1 - A_5 except that it has a larger power for controlling the armature current of the motor*. The tacho-generator and motor have independent excitation.

The purpose of this control unit is to make the speed of the motor shaft (and the rotor of the commutator via the reduction gears) proportional to the control voltage U_c of the delay device. In the operation of the speed servo system, the d.c. amplifier A_6 receives a mis-match voltage equal to the difference between the control voltage and the tacho-generator voltage. After amplification by A_6, this mis-match signal controls the voltage on the armature of the motor by varying the motor speed in such a way that the existing mis-match is eliminated.

The error in the reproduction of the input signal in this particular device can be determined by formula (3-10).

* Actually, a soviet SL-161-type and tacho-generator are used and UPT-12 type amplifiers.

The delay ranges between 0.07 and 20 sec for a delay device with 30 storage capacitors ($N = 30$) and $\omega\tau = 2\pi$ for two per cent error in signal reproduction.

Compared with the circuit of Fig.17, that of Fig.19 has a larger value $\omega\tau$, which permits reproduction of input signals at frequencies from 0 to 5 c/s.

The range for which linear reproduction of the input signal can be obtained is ± 100 V (it is determined by the range of voltages of the operational amplifiers which are used). From test results, use of cathode followers in the voltage sampling system is of great importance in the correct operation of the delay device illustrated in Fig.19. The total error associated with the discharge of the storage capacitors is no more than one per cent if use is made of 6N9S-type valves, 600 V storage capacitors (KBGM-type) of 1 μF, capacitance and if the capacitance of C_0 is 0.01 μ F.

A special commutator is used in this delay circuit (see Fig.21). Its mounting plate *1* is made of organic glass to prevent any appreciable leakage from the storage capacitors; periodically it is necessary to service and clean the contact system of the commutator.

Figure 22 shows the fundamental circuit of a delay device using the principle of sequential transmission of the input voltage along a chain of storage cells [6]. The storage unit is a chain of series-connected storage cells made in the form of cathode followers (V_1, ..., V_n) with storage capacitors at the input; all the cells are interconnected by contacts K_1, ..., K_n, the even-numbered contact sets being normally closed, and the odd sets normally open. Figure 22 also shows the circuits of the unit for controlling the switching rate of the contacts (the frequency modulator with the values V'_1, V'_2, V'_3), the device for the piecewise-linear approximation (the valves V'_4, V'_5, V'_6, V'_7) and the d.c. operational amplifiers A_2 and A_3. The d.c. operational amplifier A_1 is the output element of the delay device. It is also used to adjust for the required

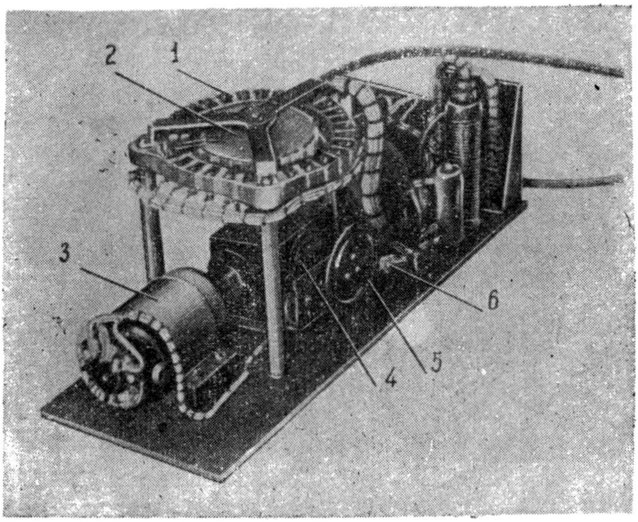

Fig. 21. Photograph of the commutator of a controlled-delay device:

1 – plate of commutator made from organic glass; *2* – rotating segments of the commutator; *3* – an SL-161 type electric motor; *4* – reduction gears; *5* – disk with saw-tooth slots (the photo-cell PC_1 covered by this disk); *6* – incandescent lamp.

gain (K=1) and to add the output signal voltage with the linear increment voltage formed at the output of the approximation circuit (A_3). This amplifier can also be used as an adder in the synthesis of a specified transfer function [12]. In that event the output voltage from all the cathode followers V_1, ..., V_n of the storage cells are fed across appropriately selected resistors which ensure the necessary scale coefficients for summation, to the input of the amplifier A_1 (similar to the connexion of the resistance R_1 to the output of the n-th cell).

In operation, the necessary magnitude of the control voltage U_c which ensures that the contact sets K_1, ..., K_n of the storage cells are switched with the required frequency, is produced at the input to the circuit for controlling the amount of time lag. Once all conditions have

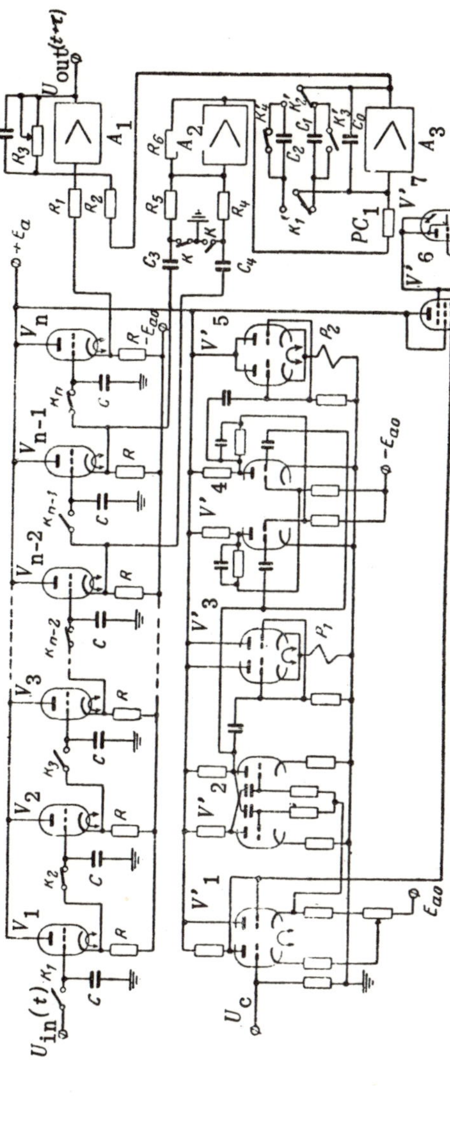

Fig. 22. Circuit diagram of a controlled-delay device with sequential transmission of the signal along a chain of storage cells. (Symbols as for Fig.19. Explanation in text).

been established, the required time lag is obtained. The electronic apparatus of the control unit is comparatively simple. It consists of a frequency-controlled multivibrator V'_2, an output cathode follower V'_3 and a d.c. input pre-amplifier (the first half of the valve V'_1). The other half of V'_1 is used as a cathode follower and serves to inter-match the steady state mode of the multivibrator V'_2 and the amplifier, and to define the carrier frequency of the multivibrator. The multivibrator generates voltage pulses at the frequency f_0 when no input signal U_c is present.

The application of the input signal U_c causes a deviation in the frequency of the multivibrator within prescribed limits $\pm \Delta f$.

The contact sets K'_1, K'_2, K'_3 and K'_4 are switched by the relay P_2 at a frequency $f/2$; for dividing the frequency f of the unit in order to control the time delay, use is made of a flip-flop circuit with anode-grid couplings using the valve V'_4. The output of the frequency divider is the cathode follower V'_5 with the relay P_2 connected in its cathode circuit.

The time lag τ can be varied "smoothly" by varying the grid bias of the multivibrator with variation of the control voltage U_c, but large changes in the ranges of time lag τ are produced by switching various resistors and capacitors in the grid circuits of the multivibrator (not shown in the diagram).

The unit for piecewise-linear approximation of the output voltage in the circuit of Fig.22 is different from that considered previously in respect of Fig.19.

The linear increment is obtained by integrating the difference between two adjacent steps of the function; here the time constant of the integrating device A_3 must be directly proportional to Δt (see equation (3-14)), or inversely proportional to the switching rate f of the unit for controlling the time lag. In the circuit of Fig.22 a change in the control voltage U_c leads simultaneously with a change in the switching rate f to a change in the voltage on the control grid of the pentode V'_6, which is a variable resistance in the circuit of the cathode ray indicator V'_7 (a 6E5S-type valve) operating as a diode, and to a change in the brilliance of the screen of this tube. This luminescence is received by the photo-cell (an FSK-1 type), which is the variable resistance of the integrating device A_3*

Thus, in regulating the time lag and changing the switching rate of the control unit, a corresponding change takes place in the time constant of the integrating device which ensures that linear increments of voltage are added to the stepped shape of the output voltage curve of the delay device.

The difference between two adjacent values of the steps of the function in the circuit of Fig.22 is obtained by capacitances C_3 and C_4 which are switched by the switches K at the frequency f of the control unit, and by the operational adding amplifier A_2. The work of each of the two input channels (C_3R_5 and C_3R_4) of this amplifier is the same, but it is in phase opposition with the other channel. At the instant represented by the circuit of Fig.22, the capacitance C_3 is charged up to the magnitude of the voltage at the output of the (n-1)-th storage cell of the delay device. In the next cycle a voltage appears at the input to the resistor R_5 after the opening of the switch K which is equal in magnitude to the difference between two adjacent "steps" of the voltage at the output of the (n-1)-th storage cell.

To avoid a considerable discharge of the capacitor C_3 through the resistor R_5, it is arranged for this resistance to be of the order of several tens of megohms.

The operation of the C_4R_4 channel of the operational

* To tune the approximation unit correctly, additional resistors are connected in parallel and in series with this photo-cell, but they are not shown in the diagram.

amplifier A_2 is similar except that it operates in phase opposition to C_3R_5. The amplifier A_2 thus adds the voltages of the difference between adjacent values of the steps of the function which are received alternately at each of its inputs. The amplifier A_2 is also used as an inverter to match the sign of the linear increment with that of the stepped output voltage of the chain of storage cells*.

Figure 23 shows curves which illustrate the performance of the approximation circuit of the delay device. That of Fig.23, *a* represents two levels of voltage (the step function f_{n-2}); the passage of this curve through the last three storage cells in sequence is shown in Fig.23, *b* and *c*. The time T corresponds to the operating period of the control unit.

Figure 23 illustrates the formation of the piecewise-linear function at the output of the approximation unit and the piecewise-linear shape of the output voltage curve of the delay device. Figure 23, *d* shows the shape of the voltage at the input to the adding amplifier A_2 along the channel C_4R_4; Fig.23, *e* refers to the channel C_3R_5. Figure 23, *f* shows the shape of the output voltage of A_2, and Fig.23, *g* the linear increment voltage as obtained after the integrating amplifier A_3. The addition of the output voltages of the cathode follower (Fig.23, *c*) and of the integrating amplifier A_3 (Fig.23, *g*) by the amplifier A_1 ensures the formation of the required piecewise-linear curve for the output voltage of the delay device (Fig.23, *h*).

The role of A_2 as an inverter will be seen from Fig. 23, *f*. This is necessary to match the sign of the linear increment with that of the stepped curve for the output voltage f_{n-2}. It follows from Fig.23, *e* and *f* that to prepare the circuit for the formation of the linear increment in the next cycle, it is necessary to provide for the

---

*As is known, a d.c. operational amplifier inverts the sign of the input voltage. The series connexion of two such amplifiers (A_2 and A_3) in the circuit of Fig.22, therefore, provides the required sign of the increment.

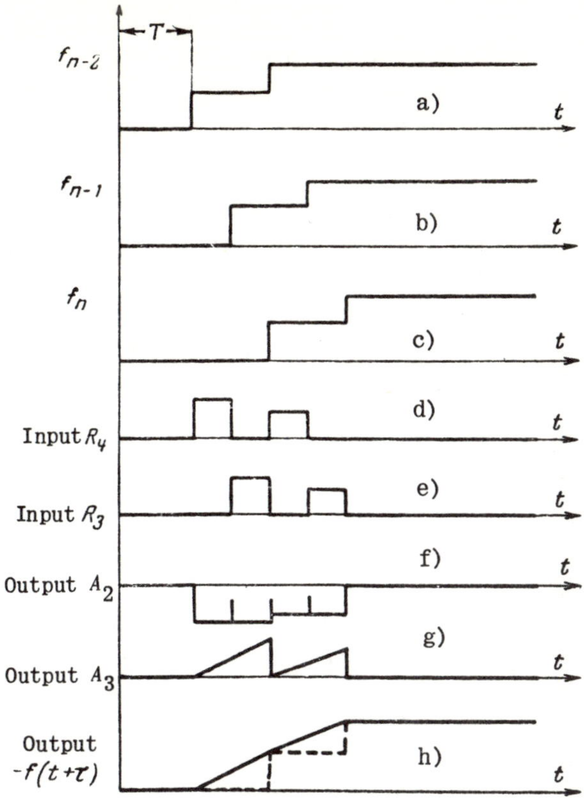

Fig. 23. Diagrams illustrating the operation of the approximation unit.

periodic discharge and switching of the capacitors C_1 and C_2 in the feedback circuit of the integrating amplifier A_3 at the frequency $f/2$. This switching is effected by the relay P_2 with the aid of the contact sets K'_1, K'_2, K'_3 and K'_4. The capacitor C_0 in the feedback circuit of the amplifier A_3, which is much smaller than C_1 and C_2, is necessary in order to reduce the error associated with the opening of the feedback circuit of the amplifier at the instant of commutation.

This delay system has $N = 24$ storage cells and

Fig. 24. Oscillograms of the input and output voltages of a controlled-delay device.

1 - input voltage; *2* - output voltage without approximation; *3* - output voltage with approximation.

$\omega\tau = 4$. The time lag can be varied between 0.2 and 20 sec and has three switched ranges. Figure 24 shows an oscillogram representing a sinusoidal input function at the frequency 0.67 c/s (*1*) and a stepped function (*2*) $U_{out}(t+\tau)$ for $\tau = 0.35$ sec. The approximated sinusoidal curve (*3*) for the output voltage of the delay device can be seen on the right-hand section of the oscillogram after the connexion of the circuit for the piecewise-linear approximation. The error in the approximation of the output function is no greater than 2.5 to 3 per cent.

A photograph of this delay device is given in Fig. 25. On the control panel the knob *1* on the left is for changing the range of τ and that on the right *2* is for continuously varying τ within this range when the control voltage U_c within the delay device has been set. The small neon lamp *3* allows the switching rate of the control unit to be observed.

At the bottom of the panel on the left there are the sockets for the input voltage $U_{in}(t)$ *4* and the control voltage U_c *5*, and on the right the socket for the output voltage $U_{out}(t+\tau)$ *6*.

The left-hand toggle switch *7* enables the control signals U_c to be obtained from the inside or from the outside of the delay device, whilst that on the right *8* is operated with

Fig. 25. Photograph of a controlled-delay device with sequential transmission of the input signal along a chain of storage cells (explanation in text)

zero setting of the operational amplifiers output and of the cathode followers of the storage cells.

Figure 26 shows a circuit which, if used, will double the accuracy with which the output voltage curve can be reproduced. Each storage cell has a cathode follower with two capacitors at its input which are switched by two contacts, namely K_1 the recording contact and K_2 the sampling contact. The capacitor C_0 is required in order to improve the operation of the system at the instant the contact K_2 is switched. The resistance R_0 serves to limit the grid currents.

During the operation of the control unit of the delay device the contacts K_1 and K_2 of each storage cell in the system (Fig.26) are switched from one position to another; when one of the storage capacitors is being charged through contact K_1, the voltage stored previously on the other capacitor is being sampled through K_2. This switching of the contacts results in the charging and sampling of the voltage of two storage capacitors after each cycle of operation of the control unit.

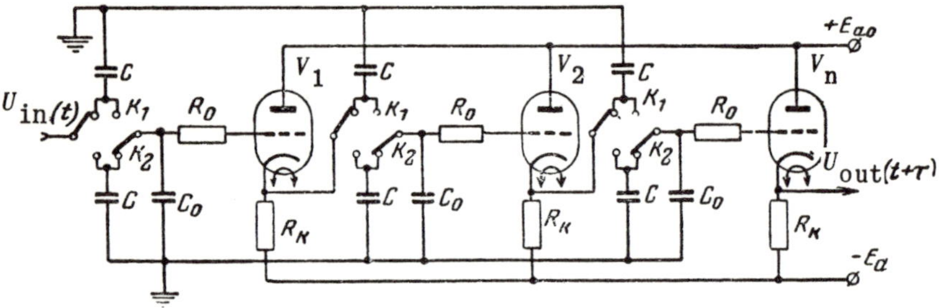

Fig. 26. Circuit diagram of an improved storage unit for a controlled-delay device with sequential transmission of the input signal along a chain of storage cells.

This system increases the accuracy with which the input signal of the delay device is reproduced by increasing the number of storage capacitors without increasing the number of cathode followers. In addition, the approximation and control units of the delay device are simplified inasmuch as there is no need to halve the frequency of the control unit. Therefore, the flip-flop circuit with the anode-grid couplings using the valve V'_4, the output cathode follower V'_5 and the relay P_2 can all be eliminated from the circuit of Fig. 22. The contact sets K'_1, K'_2, K'_3 and K'_4 must now be controlled by the common relay P_1 of the control unit.

There is no change in the connexion of the integrating operational amplifier A_3 or the correction of its time constant by the circuit consisting of valves V'_6, V'_7 and the photo-cells PC_1.

The circuits of the storage units of Figs. 22 and 26 have a number of shortcomings associated with the voltage drop along the chain of storage cells, since the gain of each cathode follower is less than unity. In addition, there is distortion of the transmitted magnitude of voltage due to the grid currents of the valves of the cathode followers.

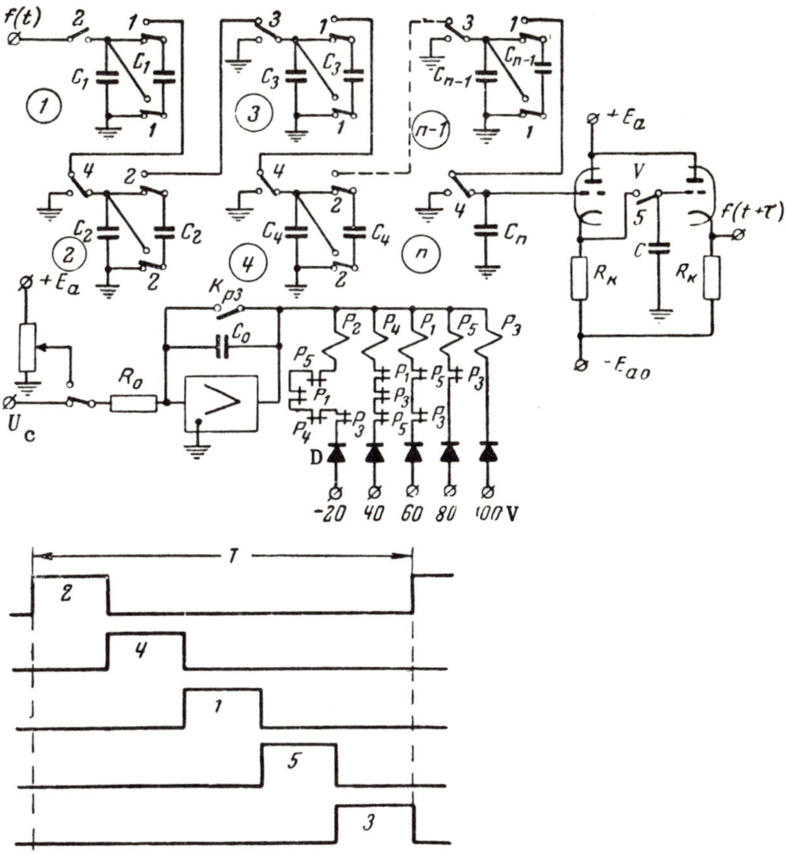

Fig. 27. Circuit diagram and cyclic operation diagram for a controlled-delay device using passive elements.

An interesting storage device system is illustrated in Fig.27. Each of the storage cells employs passive elements and so the input voltage is not attenuated during its propagation, i.e. the transfer constant of each cell is not less than unity.

The idea behind this system is that in transferring the voltage from one cell to another, twice this voltage is produced at first and it is reduced to half later. This idea can be realized in practice by simultaneously charging two storage capacitors (C_i) in each cell. In transferring voltage from this cell, the capacitors C_i are connected in sequence in such a way that the total transferable voltage is doubled in value. The total capacitance of the next $(i+1)$-th cell must be $C_i/2$ in order that the required value of voltage should be transferred. In principle it is even possible to obtain an increase in the transferred voltage by inter-relating the values of the capacitances of adjacent cells in an appropriate manner.

The voltage can be transmitted along this chain of "passive" storage cells by appropriately switching the storage capacitors of each cell and in the same way between the cells themselves. The time lag is altered by changing the switching rate of the control unit contacts.

The control of a chain of passive storage cells requires a more complicated device to regulate the time lag τ than the frequency modulator circuit used in the delay device of Fig. 22.

In fact, the output voltage from such a type of cell cannot possibly be transferred direct because at the instant it is transferred it may have changed by as much as twice its initial value at the instant of storage; each cell through which the voltage is transferred must first be discharged.

To transfer the voltage from the output of a chain of passive storage cells, use is made of the special circuit illustrated in Fig. 27 which consists of two storage cells with cathode followers. The capacitor C_n and the first half of V serve as the n-th storage cell, and the capacitor C with the other half of V are used for transferring the output voltage at those instants when it corresponds to its true (not doubled) value. This transfer is effected by switching the contact *5*. The capacitors of the passive storage cells are preliminarily discharged by the contact sets *4* and *3* in turn, the first set forming all the even cells and the second forming the odd cells.

The switching of these contacts is explained by the cycle operation diagram shown in Fig. 27. This procedure is effected by the control unit of the delay device. In the diagram T is the period and the various numbers stand for the respective contact sets in the delay device circuit.

The control unit of this system is a saw-tooth voltage generator using a d.c. integrating operational amplifier, the capacitor C_0 in the feedback circuit of which is short circuited after the operation of the relay P_3 at the output of the integrator.

The time lag τ is regulated by varying the voltage U_c at the input to the integrator and also by altering the time constant R_0C_0 (a "coarse" change in the range of τ).

An increasing voltage at the output of the integrator is accompanied in turn with the operation of the relays P_2, P_4, P_1, P_5 and P_3 which results in the opening of the preceding relays and the required switching of the cells of the storage unit.

The operation of this system will now be considered. At first, when the contacts P_2 are closed, all the even storage cells are converted for transferring the voltage, but all the odd cells are still in their original state. The next cycle entails the discharge of all the capacitors of the even cells, which results in the closing of the contact sets denoted by *4*. All the sets denoted by *1* are then switched and the voltage is transmitted from the odd-numbered cells to the storage capacitors of the even cells, after which the capacitors of the odd cells must discharge in the next cycle (contacts No. *3*). But before the discharge of the odd cells, the voltage is transferred from the output of the chain of passive storage cells. This is achieved by switching the contact set No. *5*.

This process of commutation is then repeated ad infinitum.

The output voltage curve $f(t+\tau)$ of the delay device can be approximated in the same way as before.

The technical characteristics of capacitor controlled-delay devices are given in the following Table.

It is advisable to use these capacitor controlled-delay devices for reproducing functions with a lagging variable argument at comparatively low input signal frequencies when short delays are required.

In conclusion, the design of controlled-delay devices in which the amplitude-frequency characteristic is repro-

Type of capacitor controlled-delay device	Maximum values		Maximum error	Frequency of input signals	Voltage scale
	τ	$\omega\tau$			
Device with storage capacitors and a special commutator	from 0.7 to 20 sec	6	± 2%	up to 5 c/s	± 100 V
Device with a chain of storage capacitors	from 0.1 to 20 sec	up to 3-4	up to ± 3%	up to 2 c/s	± 1000V
Device with storage capacitors and a step selector (without the approximation unit)	from 0.1 to 20 sec	up to 0.6	3-5%	up to 0.2 c/s	± 100 V

duced exactly, but the phase-frequency characteristic only approximately, will be considered. Here the idea is to express transfer function $e^{-p\tau}$ of the device as the ratio of two polynomials by means of a Pad series [15]:

$$W(p) = e^{-p\tau} \approx \frac{\tau^2 p^2 - 6\tau p + 12}{\tau^2 p^2 + 6\tau p + 12} = 1 - \frac{12 p\tau}{p^2\tau^2 + 6p\tau + 12}. \quad (3\text{-}16)$$

The amplitude-frequency characteristic is given by the expression

$$|W(j\omega)| = 1, \quad (3\text{-}17)$$

and the phase-frequency characteristic is written as

$$\varphi = -\omega\tau + 2\,\mathrm{arc\ tan}\frac{6\omega\tau}{12 - (\omega\tau)^2}. \quad (3\text{-}18)$$

Here the second term defines the error in the reproduction of the phase characteristic as compared with an ideal delay device; here $(\omega\tau) \leqslant 2.3$ for a phase error of 4°.

The transfer function (3-16) is usually realized by the operational amplifiers of the electronic analogue. A typical system is illustrated in Fig. 28.

The time lag τ is varied in steps by switching in alternate values of the capacitances C_4 and C_3 ("coarse" variation of the ranges of the time lag).

Continuous control over τ is exercised by varying the photo-cells PC_1 and PC_2 which are connected as voltage dividers at the outputs of the operational amplifiers A_1 and A_4. This system of control is similar to that described previously for the time constant of the integrating amplifier of the piecewise-linear approximation unit (Fig. 22).

A change in the control voltage U_c leads to a change in the voltage on the control grids of the pentodes 6Zh4, which are variable resistances in the circuits of the cathode-ray indicators (6E5S) operating as diodes, and to variation in the brilliance of these tubes. The change in the brilliance on these screens causes the required change in the photo-cells PC_1 and PC_2.

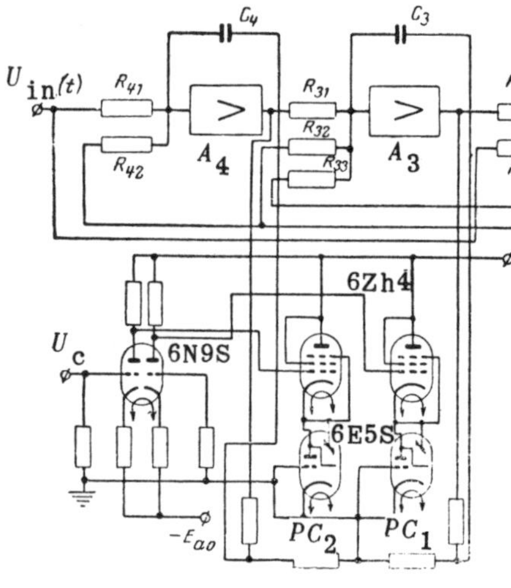

Fig. 28. Circuit diagram of a controlled-delay device based on the approximation of the function $e^{-p\tau}$ by a Pad series.

Incidentally, the transfer function (3-16) can be realized by more "economical" devices than operational amplifiers [2], but this topic is outside the scope of this book.

If necessary, the range of permissible values of ωt can be widened for this type of controlled-delay device by connecting them in series.

References

1. KORN, G and KORN, T., *Electronic analogue devices,* Foreign Literature Publishing House (1955).

2. KOGAN, B.Ya., *Electronic analogue devices and their use in the study of automatic control systems, (Elektronnyye modeliruyushchiye ustroistva i ikh primeneniye dlya issledovaniya sistem automaticheskogo regulirovaniya),* Fizmatgiz (1959).

3. FEL'DBAUM, A.A., *Electrical automatic control systems, (Elektricheskiye sistemy avtomaticheskogo regulirovaniya),* Oborongiz (1954).

4. IVANOV, V.A., Delay devices using magnetic recording, Avtomatika i telemekhanika, XVII, No.4 (1956).

5. BASOV, Ye.P., *A Function Generating Device, (Ustroistvo dlya vsoproizvedeniya funktsii).* Auth.cert.No.115944, claim No.553888, 26 June 1956.

6. DOGANOVSKII, S.A., *A controlled-delay device, (Blok reguliruyemogo zapazdyvaniya).* Auth. claim No.617454/26, 24 January 1959.

7. RENKAUF, D.C., Simulating transport lags with magnetic tape, Cont. Eng., 4 June (1957).

8. JANSSEN, J.M., Sampling delay lines, Nature, 169 (1952).

9. STONE, R.A. and DAUDL, R.A., A variable function delay for analogue computers, Trans. IRE on EC-6, No.3, September 1957.

10. GUROV, V.A., A device for producing a constant delay, (Ustroistvo dlya sozdaniya postoyannogo zapazdyvaniya). Proc. 2nd All-Union Conference on Automatic Control Theory, vol. III, Akad. Nauk SSSR (1955).

11. GRINYA, Ya.I. and KOPAI-GORA, P.N., A delay device using computing amplifiers and capacitors, Avtomatika i telemekhanika, XVII, No.6 (1956).

12. TSYPKIN, Ya.Z., *The Theory of Pulse Systems, (Teoriya impul'snykh sistem).* Fizmatgiz (1958).

13. MOLYNEUX, L., Recording low frequency phenomena on magnetic tape, Electronic engineering, December (1955).

14. DONDALSON, P.E.K., A magnetic tape recorder for electrophysiology, Electronic engineering, December (1955).

15. MORRILL, C.D., A sub-audio time delay circuit, Trans. IRE, EC-3, No. 2, June (1954).

SHORT LIST OF SYMBOLS*

U_d - voltage on capacitor C_2 (see formula 2.1) (or diode voltage)

U_{in} - input voltage

U_{out} - output voltage

U_{do} - the voltage U_d of the diode integrator at the frequency f_o. (see Fig. 3)

E - magnitude of input voltage

E_a - anode supply voltage

E_b - bias voltage

T_c - C_2R_1

T_d - C_1R_1

k - in the diagrams k represents a kilo (i.e. 1000)

k_a - gain

f_o - central carrier frequency

f_{high}, f_{low} - } extreme frequencies

M - ξ/N

ν - modulation factor

τ - delay

σ - see formula 2.3

ξ - α/α_o

ε - error

ω - input signal bandwidth

ω_o - the carrier frequency

α - total angle occupied by all the segments of the commutating device

α_o - angle between brushes

Ω - brush speed

N - number of storage capacitors

* Compiled by the Translator.